Lab Workbook
Modern Cabinetmaking
Fifth Edition

by
Brian Lee Skates
Nancy Henke-Konopasek

Publisher
The Goodheart-Willcox Company, Inc.
Tinley Park, IL
www.g-w.com

Copyright © 2016
by
The Goodheart-Willcox Company, Inc.

All rights reserved. No part of this work may be reproduced, stored, or transmitted in any form or by any electronic or mechanical means, including information storage and retrieval systems, without the prior written permission of
The Goodheart-Willcox Company, Inc.

Manufactured in the United States of America.

ISBN 978-1-63126-075-9

2 3 4 5 6 7 8 9 – 16 – 20 19 18 17 16 15

The Goodheart-Willcox Company, Inc. Brand Disclaimer: Brand names, company names, and illustrations for products and services included in this text are provided for educational purposes only and do not represent or imply endorsement or recommendation by the author or the publisher.

The Goodheart-Willcox Company, Inc. Safety Notice: The reader is expressly advised to carefully read, understand, and apply all safety precautions and warnings described in this book or that might also be indicated in undertaking the activities and exercises described herein to minimize risk of personal injury or injury to others. Common sense and good judgment should also be exercised and applied to help avoid all potential hazards. The reader should always refer to the appropriate manufacturer's technical information, directions, and recommendations; then proceed with care to follow specific equipment operating instructions. The reader should understand these notices and cautions are not exhaustive.

The publisher makes no warranty or representation whatsoever, either expressed or implied, including but not limited to equipment, procedures, and applications described or referred to herein, their quality, performance, merchantability, or fitness for a particular purpose. The publisher assumes no responsibility for any changes, errors, or omissions in this book. The publisher specifically disclaims any liability whatsoever, including any direct, indirect, incidental, consequential, special, or exemplary damages resulting, in whole or in part, from the reader's use or reliance upon the information, instructions, procedures, warnings, cautions, applications, or other matter contained in this book. The publisher assumes no responsibility for the activities of the reader.

The Goodheart-Willcox Company, Inc. Internet Disclaimer: The Internet resources and listings in this Goodheart-Willcox Publisher product are provided solely as a convenience to you. These resources and listings were reviewed at the time of publication to provide you with accurate, safe, and appropriate information. Goodheart-Willcox Publisher has no control over the referenced websites and, due to the dynamic nature of the Internet, is not responsible or liable for the content, products, or performance of links to other websites or resources. Goodheart-Willcox Publisher makes no representation, either expressed or implied, regarding the content of these websites, and such references do not constitute an endorsement or recommendation of the information or content presented. It is your responsibility to take all protective measures to guard against inappropriate content, viruses, or other destructive elements.

Cover image: pics721/Shutterstock.com

Introduction

This lab workbook is designed for use with the text, *Modern Cabinetmaking*. As you complete the questions and problems in this workbook, you can review the facts and concepts presented in the text.

Modern Cabinetmaking is a comprehensive text that explains and illustrates all aspects of fine woodworking. You will study furniture styles and designs that will support an orderly approach to the design/construction process. You will learn how to prepare sketches or drawings and how to prepare a plan of procedure before beginning work.

Modern Cabinetmaking also provides you with a thorough understanding of the technology of woodworking. You will learn the "How, What, Why, and When" of selecting appropriate materials whether they are fine hardwoods, manufactured panel products, plastics, glass, or ceramics. Hardware, abrasives, adhesives, and finishing supplies are discussed in detail.

Modern Cabinetmaking provides complete coverage of processes and techniques using both hand and power tools and machines. As the book guides you through the processes, it shows you how your decisions will affect the final product. As you study and practice your techniques, you will increase your woodworking and cabinetmaking skills. You will learn to reach make-buy decisions based on your own experience and skill. You will understand how the availability of hand, portable, and stationary tools and machines influences your decisions.

Whether you work at home, in industry, or in a school, your activities should focus on your own safety, the safety of your co-workers, and the effects of the process on the environment. Always try to protect yourself and others around you against the risk of accidents and injuries.

Lab workbook chapters correspond to chapters in the text. After reading your assignment in the text, do your best to complete the questions and problems carefully and accurately. The lab workbook chapters also contain activities related to textbook chapter content. The activities range from chapter content reinforcement to real-world application. It is important in these activities to understand any safety procedures set forth by your teacher.

Section Projects are to be completed after studying the textbook sections. The Section Projects allow you to apply the ideas, theories, and concepts learned in the textbook. The Section Projects will incorporate these ideas to help you improve your hands-on techniques and skills.

This lab workbook helps to provide the foundation on which a sound, thorough knowledge of cabinetmaking is based. This lab workbook has been designed to increase your interest and understanding of the text material.

 The content of this Lab Workbook and the text correlates to Woodwork Career Alliance (WCA) skill standards. The WCA establish a benchmark to measure and recognize an individual's skills and knowledge. The WCA skill standards help ensure that students are prepared for rigorous industry standards, and provide a pathway for advancement for professional woodworkers.

Table of Contents

Chapter Review

Section 1—Industry Overview
1. Introduction to Cabinetmaking 7
2. Health and Safety 13
3. Career Opportunities 19
4. Cabinetmaking Industry Overview 33

Section 2—Design and Layout
5. Cabinetry Styles 37
6. Components of Design 45
7. Design Decisions 51
8. Human Factors 55
9. Production Decisions 59
10. Sketches, Mock-Ups, and Working Drawings 65
11. Creating Working Drawings 71
12. Measuring, Marking, and Laying Out Materials 77

Section 3—Materials
13. Wood Characteristics 87
14. Lumber and Millwork 95
15. Cabinet and Furniture Woods 101
16. Manufactured Panel Products 107
17. Veneers and Plastic Overlays 113
18. Glass and Plastic Products 121
19. Hardware 127
20. Fasteners 133
21. Ordering Materials and Supplies 141

Section 4—Machining Processes
22. Sawing with Hand and Portable Power Tools 145
23. Sawing with Stationary Power Machines 153
24. Surfacing with Hand and Portable Power Tools 163
25. Surfacing with Stationary Machines 169

26	Shaping	177
27	Drilling and Boring	185
28	Computer Numerically Controlled (CNC) Machinery	193
29	Abrasives	199
30	Using Abrasives and Sanding Machines	203
31	Adhesives	209
32	Gluing and Clamping	217
33	Bending and Laminating	223
34	Overlaying and Inlaying Veneer	229
35	Installing Plastic Laminates	237
36	Turning	241
37	Joinery	251
38	Accessories, Jigs, and Special Machines	261
39	Sharpening	267

Section 5—Cabinet Construction

40	Case Construction	273
41	Frame and Panel Components	279
42	Cabinet Supports	287
43	Doors	295
44	Drawers	303
45	Cabinet Tops and Tabletops	311
46	Kitchen Cabinets	319
47	Built-In Cabinetry and Paneling	329
48	Furniture	337

Section 6—Finishing

49	Finishing Decisions	343
50	Preparing Surfaces for Finish	351
51	Finishing Tools and Equipment	357
52	Stains, Fillers, Sealers, and Decorative Finishes	367
53	Topcoatings	375

Section Projects

Section 1—Industry Overview

1-1	Class PowerPoint Presentation	381
1-2	Mini Flip Chart of Professional Organizations	385
1-3	Identifying Shop Hazards	389

Section 2—Design and Layout

2-1	Project Management	393
2-2	Student Desk and Desk Chair Design	403
2-3	Components of Design Poster	409

Section 3—Materials
- 3-1 Wood Samples Stringer 413
- 3-2 Moisture Content Material Testing 417

Section 4—Machining Processes
- 4-1 Making a Bench Hook 423
- 4-2 Drill Press Practice 431
- 4-3 Dry Fit 437
- 4-4 Building a Machine Jig or Fixture 441
- 4-5 Wood Joints Class Project 449
- 4-6 Make a Push Stick from a Pattern 455
- 4-7 Make a Routed Sign 459
- 4-8 Make a Sanding Block 467
- 4-9 Build a Sawhorse 471
- 4-10 Sharpening a Chisel 477
- 4-11 Make a Sign with a CNC Router 481
- 4-12 Veneering 485
- 4-13 Edge Trim 489
- 4-14 Practice Applying Plastic Laminate 493

Section 5—Cabinet Construction
- 5-1 Surfacing Stock for Cabinet Face Frames 497
- 5-2 Trinket Box 501
- 5-3 Bedside Face Frame Cabinets 509
- 5-4 Wall Cabinets—Face Frame and Carcase Construction 517
- 5-5 Frameless Bedside Stand Cabinets 525
- 5-6 Frameless Wall Cabinets 533

Section 6—Finishing
- 6-1 Fixing a Dent 541
- 6-2 Finish Sample Set 545

Name _____ Date _____ Course _____

CHAPTER **1**

Introduction to Cabinetmaking

Instructions: *Carefully read Chapter 1 of the text and answer the following questions.*

1. List three examples of wood products you use every day.

   _____

   _____

   _____

_____ 2. Conclusions made about a product design before work begins are known as ____.

3. List two factors on which all design decisions are based.

   _____

_____ 4. The purpose for having a cabinet or piece of furniture is known as ____.

_____ 5. The appearance of a cabinet is known as ____.

_____ 6. Production cabinet shops often use ____ systems to create their designs.

_____ 7. An example of a type of cabinetry designed and produced based on standards would be ____.
   A. factory-produced kitchen cabinets
   B. custom cabinets
   C. Both A and B.
   D. None of the above.

_____ 8. Which of the following describes ready-to-assemble cabinets?
   A. The finished cabinet is purchased unassembled in a neatly packaged, compact carton.
   B. The consumer assembles the cabinet with special RTA fasteners.
   C. Large furniture can be moved easily because it can be disassembled and reassembled in a new location.
   D. All of the above.

9. Describe material decisions that you would need to make if you wanted to produce a cabinet.

_____
_____
_____
_____
_____
_____
_____

_____ 10. ____ is a sheet of thinly sliced wood used to cover poor quality lumber.
 A. Particleboard
 B. Plywood
 C. Veneer
 D. Fiberglass

_____ 11. ____ decisions include choosing the tools and procedures necessary to build a product in the most efficient manner.

12. What is tooling?

_____
_____
_____

_____ 13. ____ refers to cutting or removing material.

_____ 14. ____ includes all operations where material is bent into a shape using a mold.

_____ 15. ____ involves assembling or joining two materials.

_____ 16. Making a mock-up is an example of which type of activity?
 A. Processing
 B. Preprocessing
 C. Post-processing
 D. None of the above.

_____ 17. ____ includes all tasks from cutting standard stock to finishing the product.

_____ 18. Transporting, installing, and maintaining are all ____ activities.

Name _____

19. List four activities management needs to do in order to accomplish a task.

_____

_____

_____

_____

_____ 20. Which of the following statements regarding quality is *true*?
　　　　　　　　　　　　　　　A. It can be specified by the designer or the user.
　　　　　　　　　　　　　　　B. It is measured by how well a product meets a consumer's requirements and expectations.
　　　　　　　　　　　　　　　C. Standards for quality have been established and documented.
　　　　　　　　　　　　　　　D. All of the above.

Name _____

Design Your Company Layout

Imagine yourself 10 to 15 years from now starting your own cabinetmaking company. Use the space below to draw a simple floor plan, labeling where all the elements of your company would function. Most plants have an area for the office, production, receiving supplies, and shipping products. Review the chapter to place the other elements.

Name _____ Date _____ Course _____

CHAPTER 2
Health and Safety

Instructions: *Carefully read Chapter 2 of the text and answer the following questions.*

_____ 1. Accidents occur as the result of ____.
   A. hazardous conditions
   B. reading labels, safety instructions, and caution and warning signs
   C. unsafe acts
   D. Both A and C.

_____ 2. The best preparation for any cabinet work is to ____.
   A. read
   B. watch
   C. understand
   D. All of the above.

3. Why was the Occupational Safety and Health Administration (OSHA) established?

_____

_____

4. Place a check next to each statement that describes an unsafe act.

_____ A. Performing a machine operation with proper knowledge and thorough planning.

_____ B. Working while under physical or emotional stress.

_____ C. Taking breaks when frustration interferes with concentration.

_____ D. Distracting someone who is operating a machine.

_____ E. Making setups when the machine is running.

_____ F. Putting all guards in their protective positions.

_____ G. Carrying sharp tools only by the handle.

_____ H. Lifting with the back.

5. List three ways to reduce the slippery nature of concrete or wood floors.

_____

_____

_____

_____ 6. Class 1 liquids are the most dangerous because the ____ is within room temperature.

_____ 7. Store flammable liquids in ____.
   A. old plastic soft drink bottles
   B. glass jars
   C. approved steel safety cans
   D. ungrounded metal containers

8. List three examples of safe material storage.

_____

_____

_____

_____

_____

_____ 9. Machines and electrical equipment should be wired in compliance with the ____.

_____ 10. Electrical ____ can prevent you from being shocked or electrocuted.

_____ 11. Air hoses used for ____ removal have a pressure relief nozzle.

12. Name three types of fire protection devices.

_____

_____

_____

_____ 13. Fire extinguishers should be located within ____ of any work area.
   A. 25′
   B. 50′
   C. 75′
   D. 100′

_____ 14. The soda acid liquid extinguisher is effective on Class ____ fires only.
   A. A
   B. B
   C. C
   D. D

_____ 15. The carbon dioxide extinguisher is effective on Class ____ fires.
   A. A and B
   B. B and C
   C. C and D
   D. D only

Name _____

16. List and describe five types of personal protective equipment that should be worn in the shop.

_____
_____
_____
_____
_____
_____
_____
_____

_____ 17. *True or False?* Gloves should be worn while operating power machinery.

_____ 18. ____ guards prevent machines from operating while dangerous parts are exposed?
   A. Automatic
   B. Interlocking
   C. Enclosure
   D. Point-of-operation

19. Why should people who work with woodworking machinery have first-aid training?

_____
_____
_____

Match each item to the statement that describes it.

_____ 20. Device that will automatically disconnect a circuit if a problem exists.

_____ 21. Feature that keeps a tool running even when the hand is removed.

_____ 22. Controls dust in the air.

_____ 23. Details the properties and hazards of chemical products.

_____ 24. Provides two layers of insulation.

_____ 25. Wire screen placed in the neck of a can that prevents flames from getting inside the can.

A. Dust collection system
B. Safety data sheet
C. Flame arrestor
D. Trigger lock
E. Ground fault circuit interrupter (GFCI)
F. Double-insulated tool
G. Twist-lock plug

Name _____

Shop Safety Rules

Make a list safety rules to be followed in your classroom shop.

Name _____ Date _____ Course _____

CHAPTER **3**
Career Opportunities

Instructions: *Carefully read Chapter 3 of the text and answer the following questions.*

_____ 1. The cabinetmaking industry is part of the ____ wood products industry.
   A. primary
   B. secondary
   C. aftermarket
   D. decorative

_____ 2. *True or False?* Generally speaking, cabinets are now built individually and directly on site.

_____ 3. Career advancement requires you to continue to ____.
   A. increase your knowledge of the industry
   B. improve your on-the-job skills
   C. maintain a positive attitude
   D. All of the above.

4. Provide two questions that ambitious individuals in the cabinetmaking industry might ask themselves.

_____
_____
_____
_____
_____

5. What is the *Occupational Outlook Handbook*?

_____
_____
_____
_____
_____

6. List three positions generally found within the management area in a business.

_____

_____

_____

_____ 7. Positions in the design area require some ____ skills.

8. Name four ways to obtain training in the cabinetmaking industry.

_____

_____

_____

_____

_____ 9. Products that combine wood and other materials are known as ____.

_____ 10. Altering the molecular structure of coatings to reduce the impact of ultraviolet radiation on wood is an example of ____.

_____ 11. Which of the following statements about apprenticeships is *true*?
   A. It offers the opportunity to get paid while you learn.
   B. It is not necessary to have an employer sponsor.
   C. After successful completion, the apprentice becomes a master craftsperson.
   D. It is not necessary to attend classes as part of the apprenticeship.

_____ 12. Communication, math, problem-solving, and leadership skills are known as ____ skills.

13. List four resources to use to find employment.

_____

_____

_____

_____

_____ 14. Most employment begins with ____, usually done on the job.

_____ 15. *True or False?* Employers expect 100% productivity when you start a job.

_____ 16. Quality work means ____.
   A. excellent product appearance
   B. proper tool maintenance
   C. holding to a production schedule
   D. All of the above.

Name _____

_____ 17. Increased experience at the job and having a good attitude can result in _____ opportunities.

_____ 18. If you are resigning from your job, give advance notice of _____.
   A. one day
   B. two weeks
   C. one year
   D. six months

19. List four reasons an employee might be fired.

_____
_____
_____
_____

_____ 20. A(n) _____ is the potential for advancement in responsibility, job title, and pay.

_____ 21. When building a career, join a professional _____ to observe the characteristics of people who have the type of jobs you would like.

22. Identify three strategies for career advancement.

_____
_____
_____
_____
_____
_____

_____ 23. A(n) _____ is the highest-ranking corporate officer or administrator in charge of total management of an organization.

_____ 24. A(n) _____ is someone who starts a business.

_____ 25. A(n) _____ is those individuals who will buy your product.

26. What is consignment production?

_____
_____
_____
_____

Name _____

Career Investigation

Investigate a career of your choice in the field of cabinetmaking. If possible, interview someone with that career. Conduct thorough research. Sources of information include the library, the Internet, and the guidance office at your school. Find the following information:

Career: _____

Education or training needed: _____

Salary range: _____

Description of duties: _____

_____

_____

_____

_____

Hours: _____

Working conditions: _____

_____

_____

_____

_____

Health and safety concerns: _____

_____

_____

_____

_____

Advantages of this career: _____

_____

_____

_____

_____

Copyright Goodheart-Willcox Co., Inc.
May not be reproduced or posted to a publicly accessible website.

Disadvantages of this career: _____

_____

_____

_____

_____

Other comments: _____

_____

_____

_____

Name _____

Preparing a Résumé

Design your résumé in the space provided. When you have finished, carefully reread it. Ask your instructor or counselor to read it. When you are satisfied with your résumé, type it using desktop publishing software. You may wish to make copies of your résumé to submit to potential employers.

Finding Employment

Find an advertisement for a cabinetmaking job on either a job search website or in your local newspaper. Place a copy of the advertisement in the space provided. Next, write a letter of application for the job, using the space provided. Be sure to state in the letter what you can do for the company. Your letter of application will help a prospective employer to form an impression of you.

Job Advertisement

Letter of Application

Name _____

Completing a Job Application

Pretend that you are applying for the job described in the advertisement on the previous page. Complete the following job application.

Personal Information

Name _____

Address _____

Telephone _____ E-mail address: _____

Job Interest

Position for which you are applying:

Salary desired: _____

Date you can start: _____

Have you ever been employed by this company before? _____

If yes, where? _____ When? _____

Education and Training

_____
_____
_____
_____
_____
_____
_____
_____
_____
_____
_____
_____

Employment Record

_____
_____
_____
_____
_____

Copyright Goodheart-Willcox Co., Inc.
May not be reproduced or posted to a publicly accessible website.

References

List the names of three people not related to you whom you have known at least one year.

_____

_____

_____

Additional Data

Have you ever been convicted of a crime (other than traffic, game law, or other minor violations)?_____

If yes, explain the offense and circumstances regarding the conviction:_____

_____

Age: _____

Are you a US citizen?_____

If no, do you have an alien registration card or valid US work permit?_____

Non-English languages you read: _____

Speak: _____

Write:_____

Special skills, knowledge, and abilities that qualify you for the position you are seeking:_____

_____

_____

Health Information

Are you presently or have you during the last six months been under a physician's care or in a hospital?_____

Have you ever been compensated for, or do you currently have outstanding, a job-related claim?_____

If you answered yes to any of the above, please explain. _____

_____

_____

Thank you for completing this application and for your interest in employment with us. We would like to assure you that your opportunity for employment with this company will be based on your merit only, without regard to your race, religion, sex, age, national origin, or disability.

Please read carefully:

Applicant's Certification and Agreement

I certify that the facts contained in this application are true and complete to the best of my knowledge. I understand that, if employed, falsified statements on this application shall be grounds for dismissal. I authorize investigation of all statements contained in this application. I authorize the references listed above to give you any and all information concerning my previous employment and any pertinent information they may have, personal or otherwise. I release all parties from all liability for any damage that may result from furnishing this information to you.

Signature_____ Date _____

Name _____

The Job Interview

Commonly asked interview questions are listed below. Pretend that you are being interviewed. Answer these questions as you would during an interview.

Job for which you are applying: _____

Company: _____

Why do you want to work for this company? _____

_____

_____

_____

Do you think you will like this kind of work? _____ Why? _____

_____

_____

How would you describe yourself? _____

_____

_____

_____

_____

What are your best subjects in school? _____

_____

What are your worst subjects in school? _____

_____

What other jobs have you had? _____

_____

_____

_____

Have you ever been fired from a job? _____ If so, why? _____

_____

_____

_____

Copyright Goodheart-Willcox Co., Inc.
May not be reproduced or posted to a publicly accessible website.

What is your best qualification for this job? _____

_____

_____

_____

What are your future plans? _____

_____

Why should we hire you? _____

_____

_____

_____

Name _____

Entrepreneurship

Complete the following activity about entrepreneurship.

1. Define entrepreneurship in your own words.

   _____

   _____

   _____

2. Does starting your own business appeal to you? Give two reasons for your answer.

   _____

   _____

   _____

3. List at least three products or services your business could sell.

   _____

   _____

   _____

4. Briefly describe the demand for your products or services. Consider potential customers, competition, and your abilities.

   _____

   _____

   _____

   _____

5. Based on your answer to question 4, which product or service seems to offer you the best opportunity for success?

   _____

6. How much time would your business take to manage? Do you have enough time to accomplish this?

   _____

   _____

   _____

7. What do you estimate it would cost to start and operate your business? Consider equipment, supplies, advertising, and other operating costs

   _____

8. Consider your answers to the previous question. Based on your answers, list several advantages and disadvantages of becoming an entrepreneur.

_____

_____

_____

_____

_____

Name _____ Date _____ Course _____

CHAPTER **4**
Cabinetmaking Industry Overview

Instructions: *Carefully read Chapter 4 of the text and answer the following questions.*

_____ 1. Cabinetmaking is one of several ____ in the secondary wood products industry.

2. List four subsectors in the secondary wood products industry:

_____

_____

_____

_____

_____ 3. ____ is custom designed and manufactured with a high degree of quality and precision.

_____ 4. Kitchen and bath cabinets are examples of ____ cabinetry.

5. What is meant by *reshoring*?

_____

_____

_____ 6. *True or False?* Building one-of-a-kind products is known as contract cabinetmaking.

_____ 7. Which of the following statements regarding batch production is *not true*?
   A. A fixed quantity of units is produced at one time.
   B. It must be well-timed.
   C. One company is hired by another company to make products.
   D. It is a standard practice for building stock products.

_____ 8. The ____ zone includes the offices of company executives, buyers, and production engineers.
   A. active
   B. quiet
   C. support
   D. shipping

Copyright Goodheart-Willcox Co., Inc.
May not be reproduced or posted to a publicly accessible website.

9. List seven benefits that trade associations provide.

_____

_____

_____

_____

_____

_____

_____

_____ 10. The ____ has developed an extensive set of standards and a credentialing system that provides a path for individuals to document their skills.

_____ 11. Which trade organization offers a professional certification program that teaches individuals how to operate an ethical, sustainable, and profitable custom woodworking enterprise from an owner's or manager's perspective?
A. Forest Products Society
B. Woodwork Career Alliance
C. Cabinet Makers Association
D. Woodwork Institute

_____ 12. ____ provide a great opportunity to see new machinery and products.

Match each trade show with the correct description.

_____ 13. Trade event dedicated to kitchen and bath industry.

_____ 14. Woodworking professionals gather to learn how to keep their businesses productive, competitive, and cutting edge.

_____ 15. Largest trade show in the world dedicated to wood products and processing.

_____ 16. Largest trade show for the wood industry in North America.

A. International Builder's Show
B. International Wood Fair
C. Ligna
D. International Contemporary Furniture Fair
E. Association of Woodworking and Furnishings Suppliers (AWFS) Fair
F. Kitchen and Bath Industry Show (KBIS)

Name _____

Industry Apprenticeship

Assume you have gotten your first job as a cabinetmaking apprentice, and then answer the following questions.

What trade association would you join first? Why?

_____

_____

_____

_____

_____

_____

_____

What shows would you attend? Why?

_____

_____

_____

_____

_____

_____

_____

Would you order a trade journal subscription? Why or why not?

_____

_____

_____

_____

_____

_____

_____

You, the Cabinetmaking Professional

Imagine yourself in 10–20 years. You are a cabinetmaking professional. You own a cabinet shop or are the CEO of a large cabinetmaking firm. In the space below write a report describing yourself and your responsibilities. What professional organizations do you belong to and why do you think they are important? What publications do you subscribe to and why are they important? What trade shows and conferences do you attend?

Name _____ Date _____ Course _____

CHAPTER 5
Cabinetry Styles

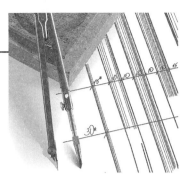

Instructions: *Carefully read Chapter 5 of the text and answer the following questions.*

_____ 1. ____ refers to the features of a cabinet that distinguish it from other pieces.

_____ 2. *True or False?* Provincial furniture has more carving than traditional pieces.

_____ 3. ____ furniture uses straight lines with very little carving.
  A. American Modern
  B. French Provincial
  C. William and Mary
  D. Contemporary

4. Place the following cabinetry styles in order, according to when they were developed: provincial, contemporary, traditional.

   _____

5. Explain how early cabinetmaking differs from modern cabinetmaking.

   _____
   _____
   _____
   _____
   _____

6. Name five traditional styles of cabinetry.

   _____
   _____
   _____
   _____
   _____

_____ 7. Identify the style of furniture shown here.

L.W. Crossan Cabinetmaker; David Gentry Photographer

_____ 8. A(n) ____ leg is a curved leg that ends with an ornamental foot.
A. contemporary
B. cabriole
C. fretwork
D. lattice

_____ 9. Identify the style of furniture shown here.

L.W. Crossan Cabinetmaker; David Gentry Photographer

_____ 10. Which of the following statements regarding Chippendale style furniture is correct?
A. It often used veneer.
B. It used inlays.
C. It often included lattice work on table aprons.
D. It had little Chinese influence.

Name _____

_____ 11. Identify the style of furniture shown here.

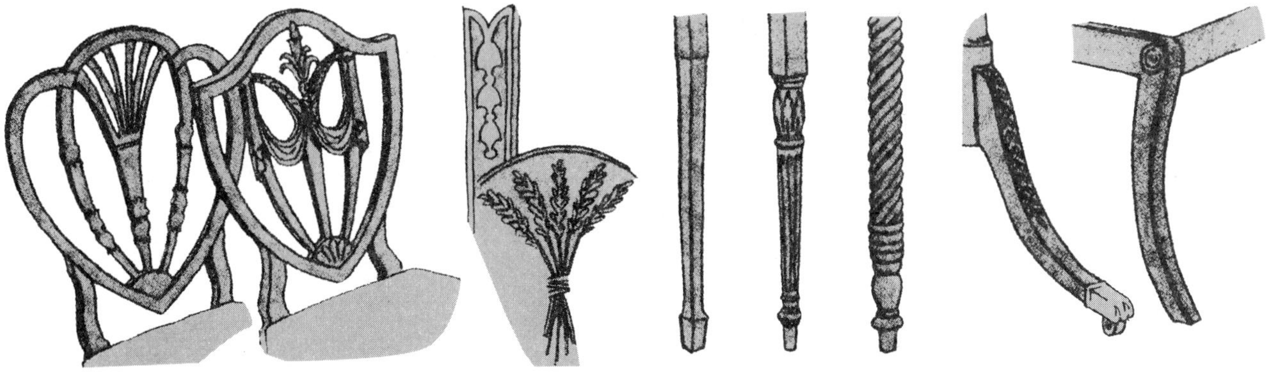

Colonial Homes

_____ 12. The secretary, a bookcase with a hinged front door, was introduced by ____.
   A. Thomas Sheraton
   B. George Hepplewhite
   C. Thomas Chippendale
   D. None of the above.

_____ 13. Simplified versions of European traditional styles were labeled ____.

_____ 14. In early America, provincial style furniture was called ____.

15. Name six provincial furniture styles.

_____

_____

_____

_____

_____

_____

Match each term with the correct description.

_____ 16. Popular from mid 1600s to about 1790 and featured graceful, curved legs.

_____ 17. Styling was crude, but some pieces were refined with European influences.

_____ 18. Cabinets were often painted with flowers or animals.

_____ 19. Extremely plain with very few decorations.

_____ 20. Chairs were finely crafted by wheelwrights.

A. Shaker
B. American Colonial
C. Windsor
D. American Modern
E. Pennsylvania Dutch
F. Duncan Phyfe
G. French Provincial

21. List five contemporary styles of cabinetry.

_____

_____

_____

_____

_____

_____ 22. Which style of furnishings combine colonial and plain styles?
   A. American Colonial
   B. American Modern
   C. Early American
   D. Shaker Modern

_____ 23. Clean, undecorated furniture is a characteristic of the ____ style.
   A. American Colonial
   B. American Modern
   C. Early American
   D. Shaker Modern

_____ 24. Which of the following types of furniture often feature an opaque lacquer surface?
   A. Shaker Modern
   B. Oriental Modern
   C. Scandinavian Modern
   D. None of the above.

_____ 25. ____ refers to how well a room matches the historic original room.

_____ 26. Coordination of furniture styles creates ____ in the interior of the home.

Identify the various styles of kitchens below.

Wood-Mode

27. _____

Name _____

Wood-Mode

28. _____

Wood-Mode

29. _____

Wood-Mode

30. _____

Wood-Mode

31. _____

Wood-Mode

32. _____

Name _____

Stylish Table and Chair

What is your favorite furniture style? Sketch a simple table and chair in that style.

Describe the style you chose and why you like it.

_____

_____

_____

_____

Name _____ Date _____ Course _____

CHAPTER 6
Components of Design

Instructions: *Carefully read Chapter 6 of the text and answer the following questions.*

1. Name the two factors that guide the development of a product during design.

   _____

   _____

2. Give three examples of functional design in cabinetry.

   _____

   _____

   _____

_____ 3. ____ refers to the style and appearance of a product.

4. Name three different ability levels among designers.

   _____

   _____

   _____

_____ 5. Lines, shapes, textures, and colors are the ____ of design.
   A. forms
   B. repetition
   C. elements
   D. intensity

_____ 6. *True or False?* Straight lines and square corners are the most difficult element to produce in cabinetry design.

7. Identify the shapes shown.

   A. _____
   B. _____
   C. _____
   D. _____
   E. _____
   F. _____
   G. _____
   H. _____
   I. _____

8. Identify each of the following images as either a primary horizontal mass or a primary vertical mass.

Thomasville Furniture Industries

A. _____   B. _____

_____ 9. ____ is the pleasing relationship of all elements in a given product design.

_____ 10. Using elements more than once to create a rhythm in a design is known as ____.

Name _____

11. Which of the following is the correct definition of formal balance?
    A. It is a relationship between height and width.
    B. It is a relationship between height and length of a product.
    C. It is equal proportions on each side of a centerline in a design.
    D. None of the above.

_____ 12. A feeling of balance in a design, even though the design is not symmetrical, is known as ____.

13. What is the value of the golden mean?

_____

_____ 14. When designing a product, the designer must consider ____.
    A. the person who will use the product
    B. the elements and principles of design
    C. convenience and flexibility
    D. All of the above.

15. Give two examples each of convenience and flexibility in cabinet design.

_____

_____

_____

_____

Match each term with the correct definition.

_____ 16. Brilliance of a color. A. Paints

_____ 17. Contain colored pigments that make an opaque finish. B. Stains

_____ 18. Red, yellow, and blue. C. Primary colors

_____ 19. The contour and feel of a product's surface. D. Value

_____ 20. The lightness or darkness of a hue. E. Intensity

_____ 21. Orange, green, and violet. F. Hue

_____ 22. Any color in its pure form. G. Secondary colors

H. Tertiary colors

I. Texture

Name _____

Principles of Design

In the space provided, mount a picture from a magazine or newspaper showing a furniture setting featuring cabinetry. Then explain how each of the principles of design is achieved.

Repetition: _____

_____

_____

Balance (formal balance): _____

_____

_____

Balance (informal balance): _____

_____

_____

Proportion: _____

_____

_____

The Golden Mean

In the space provided, draw three rectangles to the golden mean proportions. Draw one triangle with a 3″ side, one with a 2″ side, and one with a 1″ side. Round measurements to the nearest 1/16″ when calculating.

Name _____ Date _____ Course _____

CHAPTER 7
Design Decisions

Instructions: *Carefully read Chapter 7 of the text and answer the following questions.*

_____ 1. The decision-making process ____.
   A. is inflexible
   B. binds you to one design idea
   C. guides your thoughts and actions during the many steps involved in creating a cabinet design
   D. All of the above.

2. Give an example of a *need* for a cabinet and a *want* in a cabinet.

_____

_____

_____

_____

_____ 3. ____ show how space will be arranged.

4. Look at the two refined sketches below. Identify the types of sketches they represent.

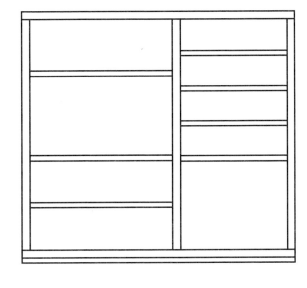 

O'Sullivan

A. _____   B. _____

_____ 5. What type of design includes details?
   A. Preliminary
   B. Final
   C. Refined
   D. Working

_____ 6. *True or False?* Dimensions are based on the intended function of a cabinet.

7. List five types of material classifications for cabinets.

_____

_____

_____

_____

_____

_____ 8. A(n) ____ contains all the materials that will be used to construct a cabinet.

Match each term with the correct description.

_____ 9. Contain all the drawings and specifications required to build a cabinet.

_____ 10. Systematic process to determine the overall cost of equipment, materials, time, and space to build a product.

_____ 11. Determines whether the product designed meets the needs of the user.

_____ 12. Provide information on materials and tools needed to build a cabinet.

_____ 13. Measures the ability of a material to withstand a load of force.

A. Strength test
B. Specifications
C. Cost analysis
D. Mock-up
E. Functional analysis
F. Make-or-buy decisions
G. Working drawings

Name _____

Cabinet Design

Identify a cabinet or piece of furniture that you would like to produce. List five questions that you would ask and answer before you produce it.

Cabinet or piece of furniture: _____

_____

Sketch of the cabinet or piece of furniture:

Questions to ask and answer: _____

_____

_____

_____

_____

_____

_____

Name _____ Date _____ Course _____

CHAPTER **8**

Human Factors

Instructions: *Carefully read Chapter 8 of the text and answer the following questions.*

1. List six human factors to consider when designing cabinetry.

   _____

   _____

   _____

   _____

   _____

_____ 2. _____ dimensions will meet the needs of a majority of people.

_____ 3. It is wise to comply with standard dimensions unless _____.
   A. you are adapting for the elderly
   B. you want your design to be unique
   C. you are adapting for the disabled
   D. Both A and C.

4. Indicate the standard dimensions for kitchen cabinets using the image provided.

   A. _____
   B. _____
   C. _____
   D. _____
   E. _____

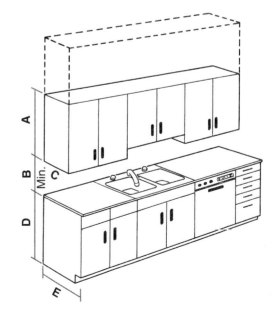

Copyright Goodheart-Willcox Co., Inc.
May not be reproduced or posted to a publicly accessible website.

_____ 5. *True or False?* Counter heights may be standard or adapted to human factors.

_____ 6. *True or False?* Machines, worktables, and walking space do not have standard dimensions associated with them.

_____ 7. What is the minimum required turning radius for wheelchairs?

_____ 8. Seat heights are measured from the floor to the top of the ____ seat cushion.

_____ 9. Which type of chair is used for dining or working?
　　　A. Lounge
　　　B. Overstuffed
　　　C. Straight
　　　D. Side

_____ 10. Standard seat height is ____.
　　　A. 17″
　　　B. 18″
　　　C. 19″
　　　D. 20″

_____ 11. The difference between table height and compressed chair seat should be ____.
　　　A. 5″
　　　B. 8″
　　　C. 11″
　　　D. 15″

_____ 12. The area that a person can see is known as ____.

_____ 13. *True or False?* Sitting restricts a person's line of sight.

14. What type of safety hazard is presented by excessive stacking of items stored overhead?

_____

Name _____

Human Factors and Safety

List two ways each of the following dangers can be eliminated or minimized.

Sharp edges and corners:

_____
_____
_____
_____
_____
_____
_____

Falling objects:

_____
_____
_____
_____
_____
_____
_____

Items within a child's reach:

_____
_____
_____
_____
_____
_____
_____

Designing for Children

In the space provided, draw cabinets similar to those shown in Figure 8-2 in the text. Change the dimensions to accommodate a 5-year-old child.

Name _____ Date _____ Course _____

CHAPTER 9
Production Decisions

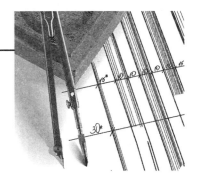

Instructions: *Carefully read Chapter 9 of the text and answer the following questions.*

1. Explain the purpose of production decisions.

   _____

   _____

   _____

2. In the diagram below, list all the essential decision-making phases in the cabinetmaking process.

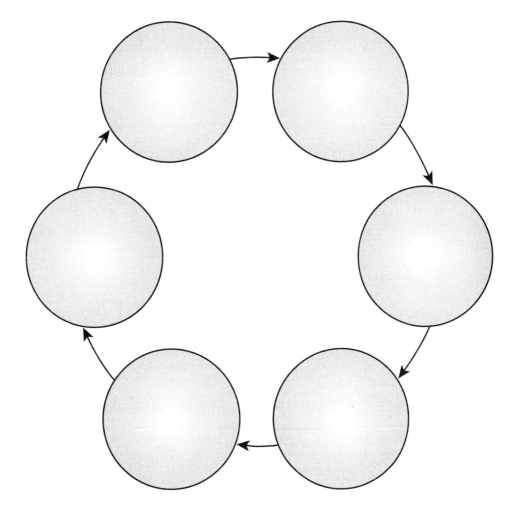

_____ 3. Having a written ____ prevents you from missing a crucial step.

4. List the 11 general steps that should be included in a plan of procedure.

_____
_____
_____
_____
_____
_____
_____
_____
_____
_____
_____

Match each term with the correct definition.

_____ 5. Accessories used to perform cutting operations.

_____ 6. Workpieces that are being processed into bill of material items.

_____ 7. A device that holds a workpiece while the operator processes it.

_____ 8. Material in its unprocessed form.

_____ 9. Space made by the blade cut of a saw.

A. Stock
B. Workpiece
C. Component
D. Tooling
E. Kerf
F. Fixture
G. Jig

_____ 10. *True or False?* The type of tools you have affects the production time and the quality of the completed cabinet.

11. Give an example of a make-or-buy decision.

_____
_____
_____

12. What two items should you be aware of when laying out workpieces?

_____
_____

Name _____

13. When cutting out materials, why should cuts be made from 1/32″ to 1/16″ (1 mm to 2 mm) oversize?

_____

_____

_____ 14. *True or False?* It is safe to assume that scales printed on machines are accurate.

15. On the mortise and tenon joint shown here, which should be cut first, the mortise or the tenon?

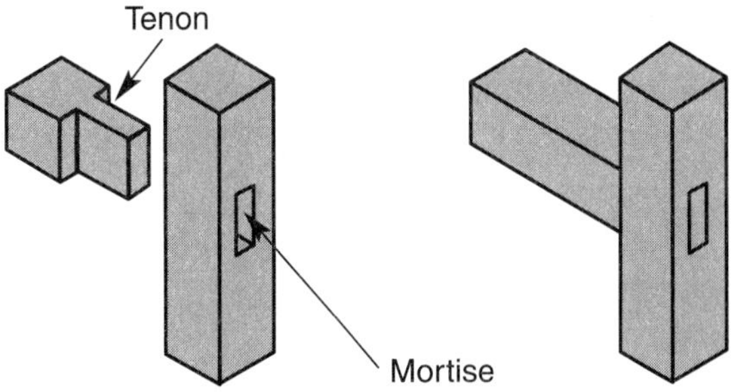

Tenon

Mortise

_____

_____ 16. Holes, such as for installing hardware, are drilled ____ the stock is squared.

_____ 17. Smoothing a surface is done ____.
    A. before cutting and shaping each workpiece
    B. after cutting, but before shaping each workpiece
    C. after cutting and shaping each workpiece
    D. None of the above.

_____ 18. What is often the most time-consuming part of the cabinetmaking process?

_____ 19. The maximum time between spreading an adhesive and joining the components is known as ____ time.
    A. cure
    B. open
    C. clamp
    D. set

_____ 20. Assembling the product without adhesive is called a(n) ____.

21. List four decisions that must be made before applying a finish to a product.

_____

_____

_____

_____ 22. *True or False?* Built-up finishes form a film on the surface of the wood.

23. List five methods for applying liquid coating materials.

_____

_____

_____

_____

_____

_____ 24. The last step in the cabinetmaking procedure is installing ____.

Name _____

Leaning Bookshelf

Make a plan of procedure for the leaning bookshelf project shown here.

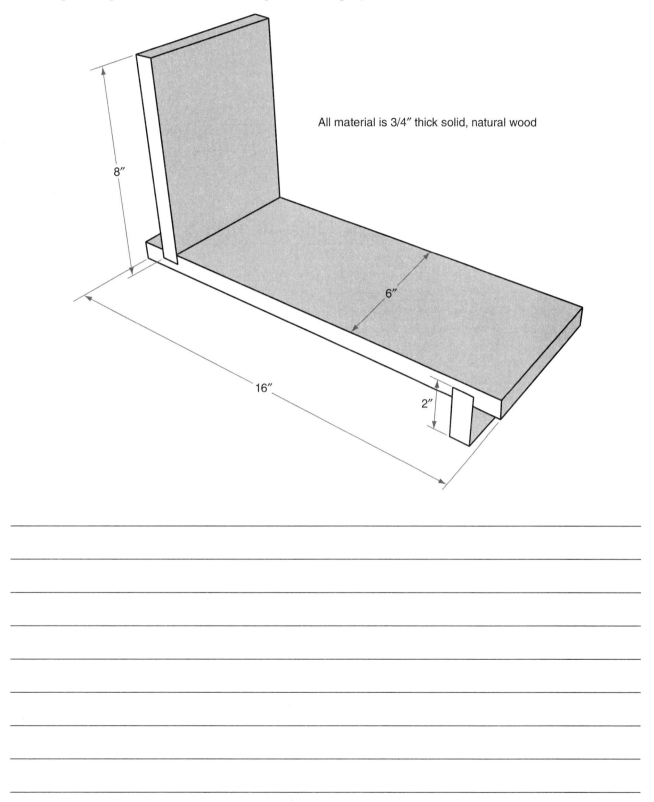

All material is 3/4″ thick solid, natural wood

Name _____ Date _____ Course _____

CHAPTER **10**
Sketches, Mock-Ups, and Working Drawings

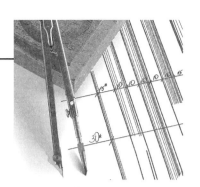

Instructions: *Carefully read Chapter 10 of the text and answer the following questions.*

_____ 1. Three-dimensional replicas of a design made of convenient, inexpensive materials are known as ____.

_____ 2. ____ communicates ideas with a drawing.

3. Identify the types of sketches shown below.

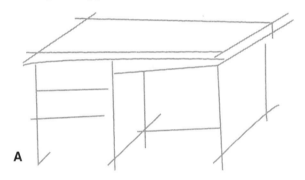

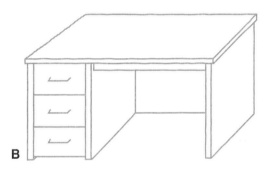

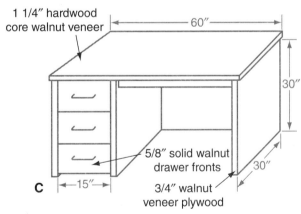

A. _____ B. _____ C. _____

_____ 4. Which sketch includes most of the information to be put on a working drawing, such as dimensions and materials?
   A. Rough
   B. Refined
   C. Thumbnail
   D. Assembly

_____ 5. A(n) ____ sketch represents a product as seen from a corner.

_____ 6. A(n) ____ sketch represents a product as if it were viewed from the front.

_____ 7. *True or False?* Isometric sketches represent a true view of an object.

8. List four skills necessary for successful sketching.

_____

_____

_____

_____

9. Describe how to create a cabinet sketch.

_____

_____

_____

_____

_____

_____

10. Describe how to create an isometric sketch.

_____

_____

_____

_____

_____

_____

_____ 11. A(n) ____ represents a circle viewed at an angle.

_____ 12. *True or False?* On perspective sketches, all horizontal lines lead to a vanishing point.

_____ 13. Working drawings include ____.
  A. views of the product showing style and features
  B. dimensions and a plan of procedure
  C. a list of materials and supplies
  D. All of the above.

Chapter 10   Sketches, Mock-Ups, and Working Drawings   **67**

Name _____

14. How are architectural drawings used?

_____

_____

_____

_____

_____ 15. Which of the following statements regarding floor plans is *true*?
  A. They explain where built-in cabinetry is to be located.
  B. They are used to determine the size and style of cabinets to be built.
  C. They represent the front view of built-in cabinetry.
  D. They indicate who should install the built-in cabinetry.

16. What information is found on material specifications?

_____

_____

_____

_____ 17. Drawings submitted for approval prior to fabrication are known as _____ drawings.

18. List the logical order to follow when reading shop drawings.

_____

_____

_____

_____

Match each term with the correct description.

_____ 19. Show the layout of the product as if it were unfolded.   A. Pictorial view
_____ 20. Allows you to see an object as if material was cut away.   B. Title block
                                                                           C. Development drawing
_____ 21. Represents a picture of the final product.   D. Detail drawing
_____ 22. Separate drawing of individual components or joints.   E. Multiview drawing
_____ 23. Often found on pictorial drawings, it includes a separate list of parts.   F. Parts balloon
                                                                           G. Section drawing
_____ 24. Shows the product disassembled.   H. Exploded view

_____ 25. A(n) _____ mock-up looks like the final product but is not functional.

Copyright Goodheart-Willcox Co., Inc.
May not be reproduced or posted to a publicly accessible website.

Name _____

Bookshelf

Make a cabinet oblique sketch, using a bookshelf of your own design.

Name _____ Date _____ Course _____

CHAPTER 11
Creating Working Drawings

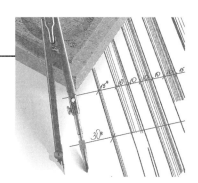

Instructions: *Carefully read Chapter 11 of the text and answer the following questions.*

_____ 1. Producing working drawings for cabinets, case goods, chairs, and tables is part of ____ technology.

2. Identify each of the following lines.

A. _____

B. _____

C. _____

D. _____

3. List the abbreviations used for each of the terms listed below.

Architectural Drawings

_____ —ceiling   _____ —minimum

_____ —flooring   _____ —maximum

_____ —between or on center   _____ —length

_____ —cabinet   _____ —height

_____ —closet   _____ —dishwasher

_____ —centerline   _____ —refrigerator

_____ —counter   _____ —vanity

Shop Drawings

_____—radius _____—finish

_____—diameter _____—laminate

_____—assembly _____—maximum

_____—average _____—minimum

_____—dimension _____—number

_____—drawing _____—round

_____—inside diameter _____—with

_____—outside diameter _____—without

Match each term with the correct description.

_____ 4. Used to draw lines vertical to or at an angle to a T-square.   A. Straightedge

_____ 5. Used for making circles and arcs.   B. Polyester film

_____ 6. Used to draw straight lines.   C. Triangle

_____ 7. Moves up and down a drafting table and works much like a T-square.   D. Parallel bar

E. Scale

_____ 8. Used to transfer distances without marking the paper.   F. Template

_____ 9. Used to draw common shapes.   G. Compass

H. Divider

_____ 10. ____ allow you to control the amount of space covered by pictorial, multiview, and detail drawings.

11. List the four essential parts of a shop drawing.

_____

_____

_____

_____

12. What types of information should appear on the title block?

_____

_____

_____

_____

Chapter 11 Creating Working Drawings 73

Name _____

_____ 13. When creating a(n) ____ drawing, the front view should be the face of the product with the most features.

_____ 14. ____ includes both linear and radial distances using extension, dimension, leader, and radius dimension lines.

15. Identify each type of dimensioning line.

A. _____
B. _____
C. _____
D. _____
E. _____

_____ 16. Lettering on shop drawings is ____.
   A. Gothic, lowercase letters
   B. Gothic, uppercase letters
   C. Gothic, both upper- and lowercase letters
   D. Times New Roman, uppercase letters

_____ 17. When drawing ____ of repeated objects, draw only one object completely and block in the others.

18. List five items of information that are included on a bill of materials.

_____
_____
_____
_____
_____

_____ 19. ____ refers to quantities of materials and supplies as you buy them.

Copyright Goodheart-Willcox Co., Inc.
May not be reproduced or posted to a publicly accessible website.

20. List five functions computers can perform in relation to the cabinetmaking process.

_____

_____

_____

_____

_____

Name _____

Alphabet of Lines

Using Figure 11-1 in the textbook, re-create and label each line in the alphabet of lines.

Lettering

Lettering is not printing or writing, it is more like drawing because specific strokes are made for each letter and number. In this activity, make five copies of each letter and numeral, using the strokes shown in Figure 11-13 in the textbook. Skip a space between each row of lettering.

Name _____ Date _____ Course _____

CHAPTER **12**

Measuring, Marking, and Laying Out Materials

Instructions: *Carefully read Chapter 12 of the text and answer the following questions.*

_____ 1. Squareness means that all corners join at a(n) ____ angle.
   A. 30°
   B. 45°
   C. 90°
   D. 180°

_____ 2. Most cabinetmakers mark workpieces with ____.

3. What procedures are taking place in the following illustrations?

Stanley Tools

A. _____   B. _____

_____ 4. The ____ is designed to make parallel lines on stock.

5. Name the two systems followed by measuring tools.

_____

_____

_____ 6. Which of the following will measure both straight lengths and curves?
   A. Flexible rule
   B. Depth gauge
   C. Caliper rule
   D. All of the above.

7. List the three purposes of squares.

___

___

___

Identify each piece of measuring equipment shown.

Lufkin Division—The Cooper Group, The L.S. Starrett Co.

8. ___

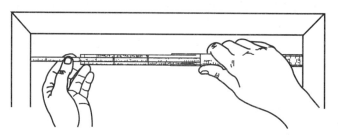

Lufkin Division—The Cooper Group, The L.S. Starrett Co.

9. ___

Patrick A. Molzahn

10. ___

Name _____

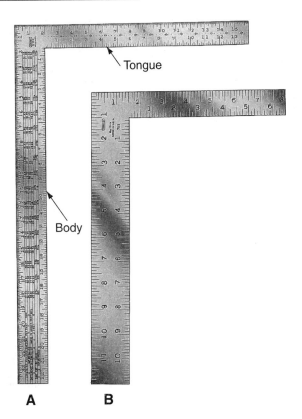

Stanley Tools

11A. _____

B. _____

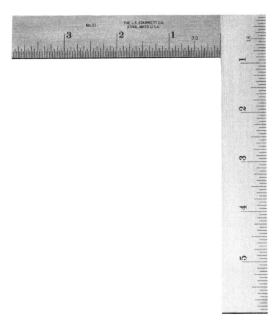

The L.S. Starrett Co.

12. _____

13. Identify the parts indicated on this combination square.

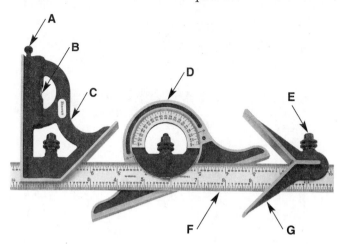

The L.S. Starrett Co.

A. _____
B. _____
C. _____
D. _____
E. _____
F. _____
G. _____

14. List three uses of a combination square.

_____
_____
_____

_____ 15. _____ tools transfer distances, angles, and contours.

Identify each piece of layout equipment shown.

16. _____

Chapter 12 Measuring, Marking, and Laying Out Materials **81**

Name _____

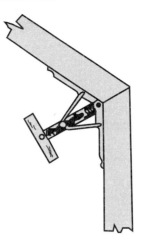

General Hardware

17. _____

The L.S. Starrett Co.

18. _____

The L.S. Starrett Co.

19. _____

The L.S. Starrett Co.

20. _____

The L.S. Starrett Co.

21. _____

Chapter 12 Measuring, Marking, and Laying Out Materials **83**

Name _____

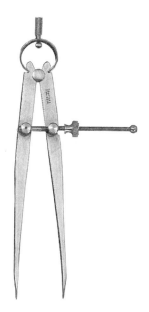

The L.S. Starrett Co.

22. _____

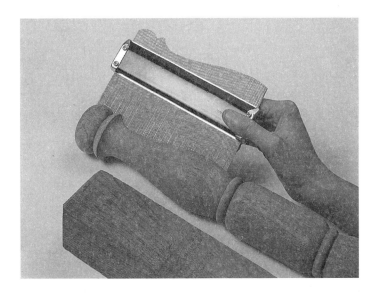

The Fine Tool Shops

23. _____

_____ 24. When marking a distance on a piece of wood, use ____.
  A. an ink marker and make a dot
  B. a pencil and make an arrow
  C. chalk and make a square
  D. a crayon and make a circle

_____ 25. Most ____ are made using a rule or square.

_____ 26. Which of the following can be used to make accurate circles and arcs?
　　A. Compasses
　　B. Dividers
　　C. Trammel points
　　D. All of the above.

27. Name two tools commonly used to lay out polygons.

_____

_____

_____ 28. A(n) ____ pattern is a way to transfer complex designs from working drawings to material.
　　A. half
　　B. detail
　　C. square grid
　　D. rectangle

_____ 29. A(n) ____ is a permanent full-size pattern.

_____ 30. *True or False?* Moving joints on measuring and layout tools can be lubricated with paste wax.

Name _____

Transferring a Design

Using the technique explained in the section "Irregular Shapes" in the text, enlarge the design in Figure 12-34. Use the following grid.

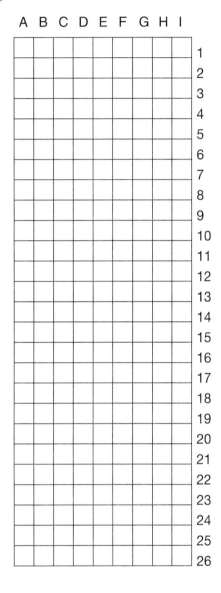

Name _____ Date _____ Course _____

CHAPTER **13**
Wood Characteristics

Instructions: *Carefully read Chapter 13 of the text and answer the following questions.*

1. List five desirable qualities of wood.

   _____
   _____
   _____
   _____
   _____

_____ 2. *True or False?* Wood is an inelastic material.

3. Identify the parts of the trees.

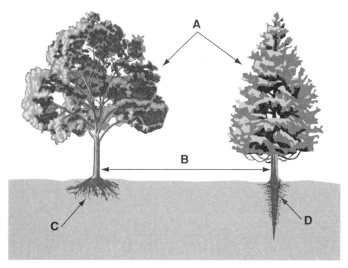

   A. _____
   B. _____
   C. _____
   D. _____

4. Identify the layers of the tree.

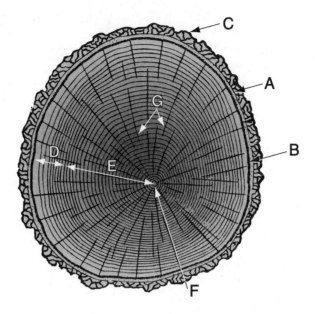

Forest Products Laboratory

A. _____
B. _____
C. _____
D. _____
E. _____
F. _____
G. _____

_____ 5. ____ are created by the growth that occurs in a single growing season.

_____ 6. The dark colored, nonliving section of a tree is called ____.
   A. heartwood
   B. sapwood
   C. earlywood
   D. latewood

_____ 7. *True or False?* Water and nutrients are carried outward from the center of a tree by wood rays.

8. Explain what a deciduous tree is and name three examples of it.

_____
_____
_____
_____

Name _____

9. Explain what a coniferous tree is and name three examples of it.

_____

_____

_____

_____

_____ 10. Wood from conifers is called ____.

11. Identify each type of wood face shown.

Forest Products Laboratory

A. _____

B. _____

C. _____

Match each term with the correct definition

_____ 12. Main passages for liquid moving from roots to crown.

_____ 13. Vertical cells in hardwood.

_____ 14. Move nutrients between the center and outer portions of a tree.

_____ 15. Substance that holds cells together.

_____ 16. Formed when space between cells expands.

A. Parenchyma cells
B. Fibers
C. Tracheids
D. Resin ducts
E. Lignin
F. Rays
G. Vessels

_____ 17. ____-porous hardwoods have similar size pores throughout the growing season.
A. Ring
B. Semiring
C. Diffuse
D. None of the above.

18. Name four factors that determine the appearance of wood.

_____

_____

_____

_____

_____ 19. The pattern of lines visible in sawn lumber that is formed by the annual rings is called ____.

_____ 20. ____ describes the amount of water in the wood cells.

21. Identify the item being used in the illustration below.

Patrick A. Molzahn

_____

Name _____

22. Provide the formula for and outline the procedure for calculating wood moisture content.

_____

_____

_____

_____

_____

_____

_____

_____ 23. *True or False?* Equilibrium moisture content does not change with the seasons.

24. Show the formula for calculating wood shrinkage.

_____

_____ 25. Wood ____ at different rates in different directions.

26. List three factors that control the weight of wood.

_____

_____

_____

27. Show the formula for calculating specific gravity.

_____

_____ 28. *True or False?* Woods with high specific gravity tend to dull tools faster.

_____ 29. ____ wood is in leaning trees of some hardwood species.

_____ 30. ____ is the capability of wood to spring back after being bent.

Name _____

Chart of Wood Characteristics

Create a chart to record the characteristics of the wood in your workshop. Using your textbook as a guide, create a list of characteristics that you wish to note. In the first column, record the name of each type of wood. In following columns, list the characteristics of that wood. For example, column headings might include color, tree classification (deciduous or coniferous), and moisture content.

The objective of this activity is to create a useful chart that you can fill out as you progress through your course and as you encounter different types of wood. Create your chart in the space provided.

Name _____ Date _____ Course _____

CHAPTER **14**
Lumber and Millwork

Instructions: *Carefully read Chapter 14 of the text and answer the following questions.*

1. List the sequence of steps used to bring wood to market as lumber.

_____

_____

_____

_____

_____ 2. ____ felling is cutting large sections of a forest at one time with heavy machinery.

_____ 3. Using ____ felling, loggers single trees are selected for harvesting.

_____ 4. What is the most common method of sawing?
   A. Plain sawing
   B. Quarter sawing
   C. Rift sawing
   D. Ripping

_____ 5. *True or False?* Quarter-sawn lumber twists and cups more than plain-sawn lumber.

_____ 6. The process of drying lumber is known as ____.

_____ 7. Wood used for cabinetmaking should be between ____ percent moisture content.
   A. 2 and 4
   B. 6 and 8
   C. 10 and 12
   D. 15 and 19

8. What is being done in the illustration below?

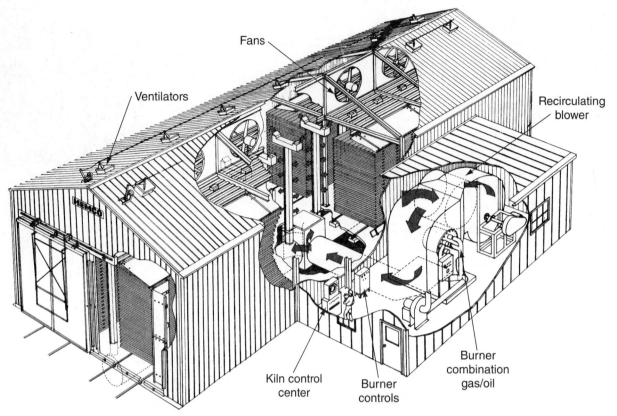

Western Wood Products Assoc.; Harvey Engineering and Manufacturing Corp.

_____

9. Name three categories of lumber defects.

_____
_____
_____

10. List four types of natural defects in wood.

_____
_____
_____
_____

_____ 11. *True or False?* Pitch pockets are openings in wood that contain solid or liquid resins.

_____ 12. Peck is caused by ____.
    A. insects
    B. fungus
    C. moisture
    D. dryness

Name _____

13. List eight types of defects caused by improper seasoning or storage.

_____

_____

_____

_____

_____

_____

_____

_____

_____ 14. Which of the following types of warp form a curve lengthwise along the face of a board from end to end?
   A. Bow
   B. Crook
   C. Cup
   D. Twist

_____ 15. ____ is the disintegration of wood fibers due to fungi.

_____ 16. Wood preservatives can be used to prevent ____.
   A. stain
   B. decay
   C. insect damage
   D. All of the above.

_____ 17. Machining defects occur most often during ____.

_____ 18. Which of the following occurs when boards are fed into the surfacer faster than the knives can cut?
   A. Torn grain
   B. Raised grain
   C. Wavy dressing
   D. Skip

_____ 19. Which of the following can cause machine burn?
   A. Dull tools
   B. Too slow of feed
   C. The cutter head rubbing in one place if the board stops during surfacing
   D. All of the above.

20. How is lumber rated by grades?

_____

_____

_____ 21. ____ grades are based on the amount of clear lumber that can be cut from a board.

22. Name three types of construction-grades lumber.

_____

_____

_____

_____ 23. ____-grade lumber is more suited to cabinetmaking than construction grade lumber.

_____ 24. ____ removes 1/8"–1/4" (3 mm to 6 mm) from the nominal (rough) size.

25. Why would a cabinetmaker specify KD lumber on an order form?

_____

_____

_____ 26. ____ consists of specialty items frequently processed from moulding-grade lumber.

_____ 27. ____-grade wood mouldings are suitable for natural or clear finishes.
  A. R
  B. P
  C. N
  D. M

28. When ordering millwork, what should be specified?

_____

_____

_____

_____

_____ 29. ____ are used as both support and decoration on stair rails.

30. Identify the following items.

A. _____
B. _____
C. _____

Name _____

Specialty Item

Brainstorm ideas for a specialty item that could be mass-produced in your class and sold to other cabinetmaking businesses as an add-on for their cabinets. Write your ideas in the space below. See Figure 14-34 in the text for inspiration and ideas.

Name _____ Date _____ Course _____

CHAPTER **15**

Cabinet and Furniture Woods

Instructions: *Carefully read Chapter 15 of the text and answer the following questions. Match each term with the correct description.*

_____ 1. Botanical name of a tree, given in Latin.

_____ 2. Wood cut from coniferous trees.

_____ 3. Growth that occurs in the spring.

_____ 4. Growth that occurs in the summer.

_____ 5. A ring caused by the addition of earlywood and latewood growth to the trunk of a tree.

_____ 6. Wormholes no defect

_____ 7. Wood that has large pores that are cut open during machining.

_____ 8. Wood that has small pores.

_____ 9. Growth that occurs across the normal grain direction.

_____ 10. The smooth or rough feel of the wood surface.

_____ 11. A measure of how likely wood is to swell or shrink when exposed to moisture.

_____ 12. A shiny appearance when sanded smooth.

_____ 13. The outer layers of a tree that carry nutrients and water.

_____ 14. Tells how easily the wood can be cut, surfaced, sanded, or processed by other means.

_____ 15. Wormholes a defect

A. Texture
B. Luster
C. Open grain
D. Density
E. Latewood
F. Sapwood
G. Annual ring
H. Hardwood
I. Dimensional stability
J. Heartwood
K. Softwood
L. Closed grain
M. Genus
N. Species
O. Machining/working qualities
P. Cross grain
Q. Earlywood
R. WHAD
S. WHND

16. Various working properties of woods are given below. Next to each property, list at least one wood that you would choose that exhibits that property.

Planing _____

Drilling _____

Sanding _____

Turning _____

Gluing _____

Nail and screw holding _____

Bending—difficult _____

Bending—easy _____

Hardness—hard _____

Hardness—soft _____

Compression strength—weak _____

Compression strength—strong _____

Shock resistance—low _____

Shock resistance—high _____

Stiffness—limber _____

Stiffness—stiff _____

Availability—lumber _____

Availability—veneer _____

Cost—expensive _____

Cost—inexpensive _____

17. List five hardwoods that require filler to make a smooth surface for finishing.

_____

_____

_____

_____

_____

Name _____

Wood Characteristics

Provide a brief description of the characteristics of each wood and list two of its uses.

Wood	Characteristics	Uses
Alder, red		
Ash		
Banak		
Basswood		
Beech		
Birch		
Butternut		
Cedar, aromatic red		
Cherry		
Chestnut		
Cottonwood		
Cypress, bald		
Ebony		
Elm, American		
Fir, Douglas		
Gum, red		

Wood	Characteristics	Uses
Hackberry		
Hickory		
Lauan		
Limba		
Mahoganies, genuine		
Mahogany, African		
Maple, hard		
Maple, soft		
Oak, red or white		
Paldao		
Pecan		
Pine, Ponderosa		
Pine, sugar		
Pine, yellow		
Primavera		
Redwood		
Rosewood		
Santos Rosewood		

Name _____

Wood	Characteristics	Uses
Sapele		
Sassafras		
Satinwood		
Spruce		
Sycamore		
Teak		
Tulip, American (Yellow Poplar)		
Walnut, American		
Willow		
Zebrawood		

What Wood to Use?

Imagine you are in the business of making rocking chairs. What wood would you use? Use this space to explain how you came to your decision and why you picked the wood you did. Include considerations such as cost, appearance, and workability, along with other design decisions.

_____

_____

_____

_____

_____

_____

_____

_____

_____

_____

_____

_____

_____

_____

_____

_____

_____

_____

_____

_____

_____

_____

_____

Name _____ Date _____ Course _____

CHAPTER **16**

Manufactured Panel Products

Instructions: *Carefully read Chapter 16 of the text and answer the following questions.*

_____ 1. Manufactured panel products are widely used by cabinetmakers to ____.
  A. create large surfaces for case goods
  B. reduce the need for edge gluing lumber to make wide boards
  C. reduce production time without sacrificing quality
  D. All of the above.

_____ 2. *True or False?* Panel products are typically less stable than solid lumber.

3. List three categories of panel products.

_____

_____

_____

_____ 4. Structural wood panels are selected when ____ are required?
  A. beauty and appearance
  B. stability and strength
  C. strength and appearance
  D. flexibility and structure

_____ 5. What structural panel is manufactured with a core material sandwiched between two pieces of face veneer?

_____ 6. The thickness of veneer-core plywood refers to the number of ____.

7. Identify the type of plywood shown and identify each item indicated on the diagram.

   _____

   A. _____
   B. _____
   C. _____
   D. _____

_____ 8. Which of the following statements regarding lumber-core plywood is false?
   A. It has a solid wood center and thin veneer faces.
   B. The core may be thin, laminated strips of wood or wider boards.
   C. The veneer layers between the core and back and front faces are called cross-bands.
   D. The cross-band grain is at 45° to the faces.

_____ 9. Particleboard core plywood is ____.
   A. more expensive than veneer-core plywood
   B. approximately the same price as veneer-core plywood
   C. more expensive than lumber-core plywood
   D. approximately the same price as lumber-core plywood

10. Identify the items indicated on the grade trademark.

   A. _____
   B. _____
   C. _____
   D. _____
   E. _____
   F. _____
   G. _____
   H. _____

_____ 11. The ____ that bonds the layers determines, in part, how the plywood is used.

Name _____

Match each term with the correct description.

_____ 12. Made of wood chip or fiber core faced with a veneer.

_____ 13. Made of wood wafers and resin adhesive.

_____ 14. Composed of small wood flakes, chips, and shavings bonded together with resins or adhesives.

_____ 15. Manufactured with strands of wood that are layered perpendicular to each other.

A. Oriented strand board
B. Composite panel
C. Structural particleboard
D. Performance-rated structural wood panels
E. Waferboard
F. Phenolic resin

_____ 16. ____ panels provide the appearance and strength of solid hardwood, yet they are much less expensive.

17. List four items that determine the appearance of face veneers of hardwood plywood.

_____

_____

_____

_____

_____ 18. Standards for hardwood plywood are set by the ____.

19. Name ten specifications to include when ordering hardwood plywood

_____

_____

_____

_____

_____

_____

_____

_____

_____

_____

_____ 20. *True or False?* Low-density fiberboard and particleboard are manufactured using heat and pressure.

_____ 21. Which of the following types of hardboard is the strongest and has good finishing qualities?
A. Standard hardboard
B. Tempered hardboard
C. Service hardboard
D. Plywood hardboard

_____ 22. Medium density fiberboard (MDF) is ____ than hardboard.

_____ 23. Because of its smooth surface, ____ is often used as substrate material for laminations on cabinets, countertops, tabletops, and drawer fronts.

_____ 24. ____ particleboard is a low-cost alternative to fiberboard.

_____ 25. The nail or screw holding ability of a panel is related to its ____.

Name _____

Choosing the Right Material for the Job

Imagine you are the owner of a cabinet shop and you want to build the best cabinets for the least amount of money. Based on your study of Chapter 16, explain what materials decisions you made for the following items.

Case parts that will be seen: _____

_____

_____

_____

_____

_____

_____

_____

Case parts that will not be seen: _____

_____

_____

_____

_____

_____

_____

_____

Countertops that will be laminated: _____

_____

_____

_____

_____

_____

_____

_____

Name _____ Date _____ Course _____

CHAPTER **17**

Veneers and Plastic Overlays

Instructions: *Carefully read Chapter 17 of the text and answer the following questions.*

_____ 1. A(n) _____ is any thin sheet material that covers a core material.

2. Name the two types of veneer.

_____

_____

_____ 3. The most common thickness of _____ veneer is 1/42".
   A. flat
   B. flexible
   C. overlay
   D. substrate

_____ 4. Logs from which veneer is cut are debarked and cut to length to form _____.

_____ 5. *True or False?* The appearance of veneer depends greatly on the _____.

_____ 6. Swinging the flitch against the knife is known as _____.
   A. rotary cutting
   B. slicing
   C. stay-log cutting
   D. chuck turning

7. Identify each of the following methods of cutting veneer.

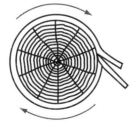 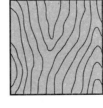

A

A. _____

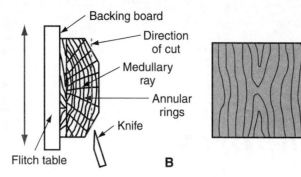

B. _____

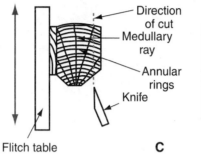
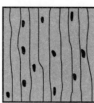

C. _____

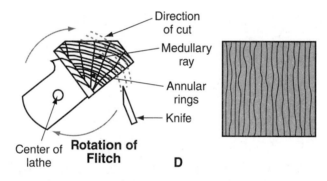

D. _____

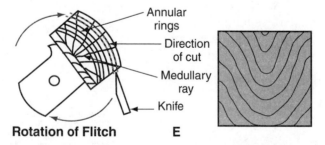

E. _____

_____ 8. The grain pattern created by most ____ is wide.
　　　　　　　　　　 A. rotary cutting
　　　　　　　　　　 B. flat slicing
　　　　　　　　　　 C. quarter-slicing
　　　　　　　　　　 D. rift cutting

_____ 9. Logs are sectioned into four flitches for ____.

Name _____

_____ 10. ____ is a method of stay-log cutting that produces a large, U-patterned grain.
   A. Flat slicing
   B. Half-round cutting
   C. Rift cutting
   D. Quarter slicing

_____ 11. Veneer sliced from ____ has a circling, wavy, knotty pattern.

_____ 12. ____ is done by splicing veneers together with the grain pattern in specific directions.

13. Describe how each veneer match is made.

   Book match: _____

   _____

   Slip match: _____

   _____

   Diamond match: _____

   _____

   Reverse diamond match: _____

   _____

   Four-way center and butt match: _____

   _____

   Vertical butt and horizontal book leaf match: _____

   _____

14. Identify each type of matching.
   A. _____
   B. _____
   C. _____
   D. _____
   E. _____
   F. _____
   G. _____

_____ 15. Veneer _____ are made by cutting veneer into a pattern and bonding it to a wood backing.

16. What is being done in the photo below?

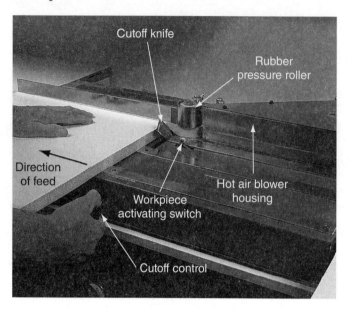

Chuck Davis Cabinets

_____

_____ 17. *True or False?* Plastic overlays can be either rigid or flexible.

18. List three uses for plastic overlays.

_____

_____

_____

19. Identify the layers of the high-pressure decorative laminate.

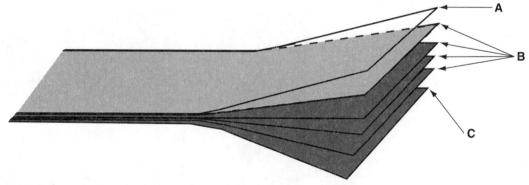

A. _____

B. _____

C. _____

Name _____

Match each type of HPDL to the correct use.

_____ 20. Provides a decorative surface in areas where there will be little wear.

_____ 21. Useful for areas that have curved corners and edges.

_____ 22. Applied to the opposite of a substrate covered by a decorative laminate.

_____ 23. Most widely used type of laminate.

_____ 24. Greater abrasion and scuff resistance, used in commercial, contract, and institutional settings.

A. General-purpose
B. VGL and VGS
C. Post-forming
D. Cabinet-liner
E. Backing sheet
F. High-wear
G. Compact laminate

_____ 25. *True or False?* The major use of LPDL panels is for cabinet exteriors.

26. Identify the parts of this flexible overlay.

A. _____

B. _____

Name _____

Material Warpage

Look at the illustration provided. It is a cross section of a piece of industrial particleboard. Laminate is glued to one side, with no backer board glued to other. Draw a similar illustration, showing how the piece will warp if placed in a humid environment. Remember that wood expands when it takes in moisture.

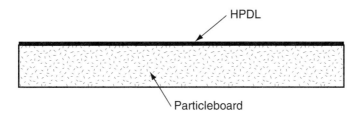

Name _____ Date _____ Course _____

CHAPTER **18**

Glass and Plastic Products

Instructions: *Carefully read Chapter 18 of the text and answer the following questions.*
Match each term with the correct description.

_____ 1. Ensures a controlled break when cutting glass.

_____ 2. Distort beyond use if reheated.

_____ 3. Can be reheated and reformed many times.

_____ 4. Process in which glass is reheated and quenched quickly to increase strength.

A. Thermoplastic materials
B. Annealing
C. Glass
D. Thermoset plastics
E. Tempered
F. Plastics

5. Name three forms glass and plastic products can take.

_____
_____
_____

6. List three uses of flat glass in cabinetry.

_____
_____
_____

_____ 7. ____ glass is thicker and often stronger than sheet glass.

_____ 8. Glass that has been worked in some way to manipulate the surface, color, or pattern of the glass is called ____ glass.
   A. flat
   B. float
   C. tinted
   D. decorative

_____ 9. ____ glass is made by adding coloring agents to molten glass.

_____ 10. *True or False?* Glass is installed before a cabinet has been assembled and finished.

_____ 11. Which of the following statements regarding glass cutting is false?
  A. The glass is actually cut.
  B. It can be done by hand or machine.
  C. The cutting wheel can be pushed or pulled across the glass surface.
  D. Lubricant is used to reduce the amount of glass-surface flaking.

_____ 12. Glass has a tendency to ____, making it very difficult to break cleanly.

_____ 13. Glass can be fractured by ____.
  A. bending glass clamped to a scoring machine
  B. bending with pliers
  C. tapping
  D. All of the above.

14. List three methods for mounting glass.

_____

_____

_____

_____ 15. ____ glass refers to panels using colorless glass.

16. Place the following steps for preparing and assembling leaded and stained glass panels in order.

_____ A. Mounting the glass in the frame

_____ B. Laying out full-size patterns

_____ C. Fitting lead came

_____ D. Cutting glass

_____ E. Grouting along the lead seams

_____ F. Soldering the lead joints

_____ 17. ____ plastic is a rigid plastic often used to replace glass in many applications.

_____ 18. Liquid ____ resin is used with fiberglass and as a casting and coating material.

_____ 19. Which of the following can be either a thermoplastic or thermoset plastic?
  A. Polystyrene
  B. Polyethylene
  C. Polyurethane
  D. None of the above.

20. Name two ways plastic can be cut.

_____

_____

Name _____

21. Why is a backer board placed under plastic when it is being drilled?

_____

_____

_____ 22. *True or False?* When working with plastics, finishing should be necessary only on the sheet edges.

_____ 23. ____ dissolve plastic causing joint surfaces to soften and the plastic to flow together.

24. Identify the decorative edge designs shown.

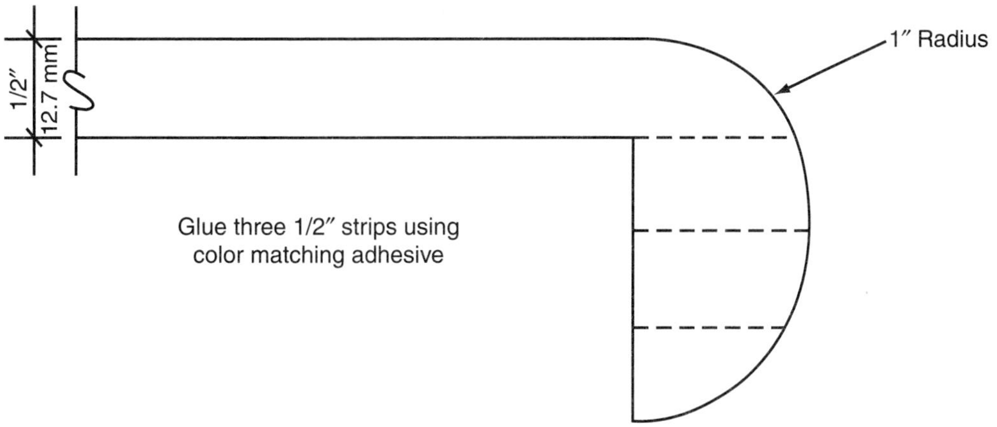

A

A. _____

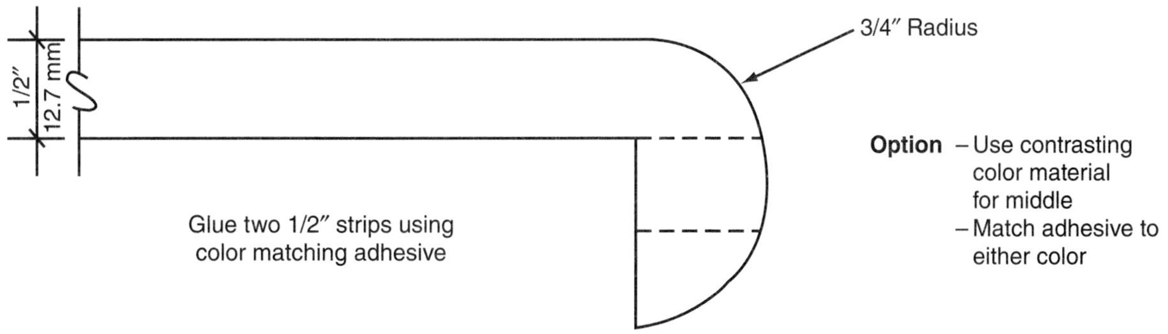

B

B. _____

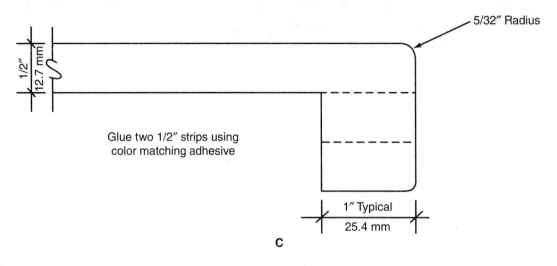

C. _____

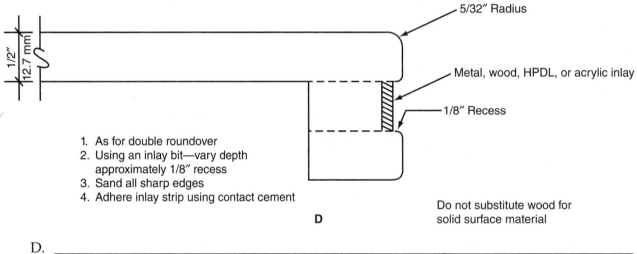

D. _____

_____ 25. The process of heating and bending materials is known as ____.

26. What steps are included in final finishing of setting solid surface material tops?

_____

_____

Name _____

Countertop Materials

You have been contracted to build a countertop for a convenience store. This countertop must withstand constant use. Draw a cross-section diagram and specify the materials you would use for the countertop.

Name _____ Date _____ Course _____

CHAPTER **19**
Hardware

Instructions: *Carefully read Chapter 19 of the text and answer the following questions.*

_____ 1. Which of the following types of pulls include a backplate that supports a hinged pull?
A. Surface mount
B. Flush mount
C. Bail
D. Ring

_____ 2. *True or False?* Upper and lower sliding door tracks must be perpendicular.

3. Identify the case front styles shown here.

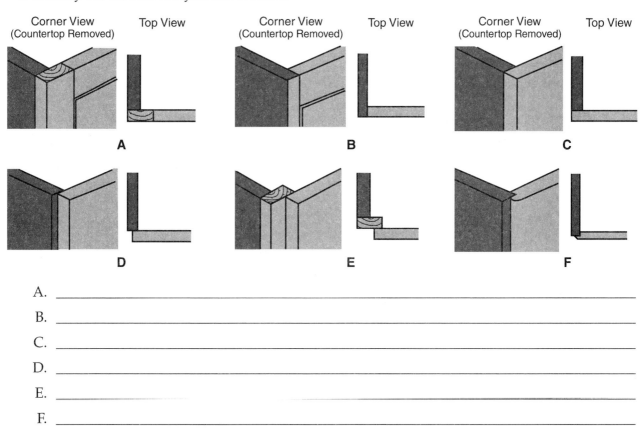

A. _____
B. _____
C. _____
D. _____
E. _____
F. _____

4. Identify each type of hinge shown and name at least one application for it.

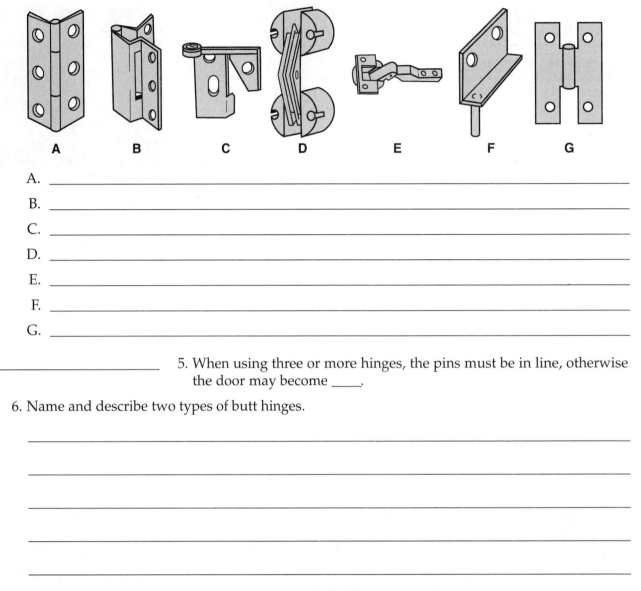

A. _____
B. _____
C. _____
D. _____
E. _____
F. _____
G. _____

_____ 5. When using three or more hinges, the pins must be in line, otherwise the door may become ____.

6. Name and describe two types of butt hinges.

_____
_____
_____
_____
_____

7. Identify the components of the European concealed hinge shown here.

A. _____
B. _____
C. _____
D. _____
E. _____
F. _____
G. _____
H. _____

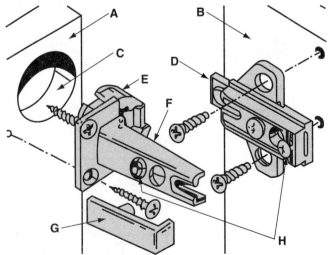

Häfele America Co.

Name _____

8. Name three glass hinge styles.

_____

_____

_____

Match each term with the correct description.

_____ 9. Consists of two spring-loaded balls with a strike between them.

_____ 10. Consists of a bent spring steel catch and a ball head screw used as a strike.

_____ 11. Hooks onto the strike and a spring-action lever is pushed to release the catch.

_____ 12. Includes a spring arm with a roller that seats in a concave strike when the door is closed.

_____ 13. Has one roller and a hook shaped strike.

A. Friction catch
B. Single roller catch
C. Bullet catch
D. Elbow catch
E. Ball catch
F. Magnetic catch
G. Spring catch

_____ 14. Doors with latches do not require ____ since the hardware opens the door as well as holds it closed.

_____ 15. ____ slides permit the entire drawer body to extend out of the drawer.
A. Standard
B. Full-extension
C. Full-extension with over travel
D. Half-extension

_____ 16. *True or False?* A drawer track is attached to the inside of a cabinet.

_____ 17. The body of a(n) ____-action lock is flat on one or both sides.
A. bolt
B. ratchet
C. cam
D. strip

_____ 18. ____ protect the bottoms and legs of freestanding cabinetry.

Name _____

Which Way to Open?

Imagine you own a cabinet shop. You wish to standardize the door style used on cabinets from your shop. Refer to Figure 19-10 in the textbook. What system would you choose? Why? Draw the door style you chose and explain why you chose it.

Door style:

Why did you choose this door style?

_____

_____

_____

_____

Name _____ Date _____ Course _____

CHAPTER **20**
Fasteners

Instructions: *Carefully read Chapter 20 of the text and answer the following questions.*

1. List four examples of fasteners.

   _____

   _____

   _____

   _____

_____ 2. ____ fasteners allow for easy assembly and disassembly.

_____ 3. *True or False?* A nail has more holding power than a screw.

4. Identify each type of nail shown.

   A. _____
   B. _____
   C. _____
   D. _____
   E. _____
   F. _____

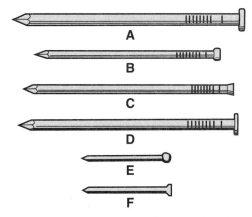

Continental Steel Corp.

_____ 5. Most cabinetmakers drive nails using a(n) ____ fastening tool.

_____ 6. Nails are measured by ____ and specific dimensions.

_____ 7. The length of a nail should be ____ time the thickness of the materials being fastened.
   A. two
   B. three
   C. four
   D. five

_____ 8. Drilling a(n) ____ hole will prevent a board from splitting when driving nails.

9. What fastening method is shown in this drawing?

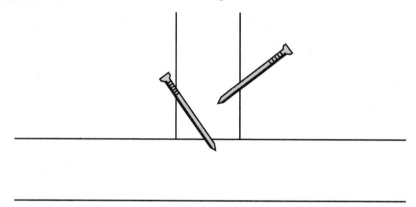

_____

10. Which staple is sized correctly?

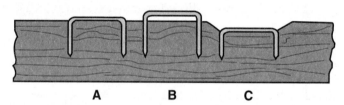

_____

Match each term with the correct description.

_____ 11. Eight-prong staples that join parts without cutting or splintering wood fibers.

_____ 12. Small brass nails with round heads, used for decorative purposes.

_____ 13. Angled fasteners

_____ 14. Installed in hidden areas, specified by crown width and leg length.

_____ 15. Shaped pieces of steel.

A. Skotch fasteners
B. Staples
C. Chevrons
D. Escutcheon pins
E. Tacks
F. Clamp nails
G. Corrugated fasteners

_____ 16. A(n) ____ is a raised, helical rib or ridge around the interior or exterior of a cylindrically shaped fastener.

Name _____

17. Identify each type of screw shown.

Liberty Hardware

A. _____

B. _____

C. _____

D. _____

E. _____

F. _____

G. _____

H. _____

I. _____

18. List the six steps for installing wood screws.

_____
_____
_____
_____
_____
_____

_____ 19. ____ may be used to drill pilot and clearance holes at the same time.

_____ 20. Flat head screws and others with tapered heads require ____.

21. Identify each type of screw shown.

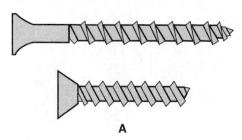
Liberty Hardware

A. _____

Liberty Hardware

B. _____

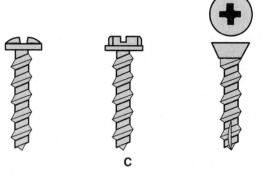
Liberty Hardware

C. _____

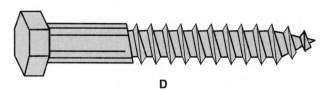

D. _____

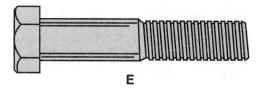
Graves-Humphreys, Inc.

E. _____

Name _____

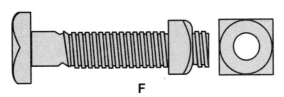
F

Graves-Humphreys, Inc.

F. _____

_____ 22. Install ____ nuts when frequent assembly and disassembly is desired.

23. Name three types of anchors.

_____

_____

_____

24. What is the purpose of repair plates?

_____

_____

_____ 25. ____ connectors install quickly and can be used where appearance is not a factor.
  A. Bolt and cam
  B. Concave bolt
  C. Wedge pin
  D. Plug and socket

_____ 26. ____ connectors are designed for joints that require less holding power.
  A. Bolt and cam
  B. Concave bolt
  C. Wedge pin
  D. Plug and socket

_____ 27. Keku fasteners are used for installing ____.
  A. wainscoting
  B. framed mirrors
  C. wall panels
  D. All of the above.

Name _____

Purchasing Agent

Most manufacturing companies have purchasing agents. These individuals are responsible for finding materials and buying those materials at a cost that keeps the company competitive.

Imagine you are the purchasing agent for a cabinetmaking company. List several fasteners that you would buy in large quantities because they are used in all the cabinets your company produces. Explain why you included those fasteners in your inventory.

_____

_____

_____

_____

_____

_____

_____

_____

_____

_____

_____

_____

_____

_____

_____

_____

_____

_____

_____

_____

_____

_____

_____

_____

_____

_____

Name _____ Date _____ Course _____

CHAPTER **21**

Ordering Materials and Supplies

Instructions: *Carefully read Chapter 21 of the text and answer the following questions.*

_____ 1. Being ____ means getting the right price for the right quantity at the right time at the quality level you need.

_____ 2. Being thorough means you must study the ____ in order to create a complete, detailed, accurate order.
  A. product views
  B. bill of materials
  C. plan of procedure
  D. All of the above.

_____ 3. Evaluating the cost of good from various vendors is known as ____ shopping.

_____ 4. *True or False?* Liquids in larger containers are generally less expensive per ounce than those in smaller containers.

_____ 5. ____ are items that become part of the finished product.

Match each term with the correct description.

_____ 6. Ordered by species and specific size.   A. Abrasives
_____ 7. Usually sold in random widths and lengths.   B. Hardwood
_____ 8. Ordered by shape or pattern number and linear foot.   C. Softwood
_____ 9. Ordered by the sheet, usually 4′ × 8′ (1220 mm × 2440 mm).   D. Fasteners
  E. Plywood
  F. Mouldings

_____ 10. Glass is sized by ____.
  A. thickness
  B. dimensions
  C. special treatment
  D. All of the above.

11. Name five specifications used when ordering fasteners.

_____

_____

_____

_____

_____ 12. ____ products include fillers, sealers, stains, and topcoatings.

_____ 13. Finishes require a compatible type of ____.

_____ 14. ____ for cabinetmaking include consumable items that do *not* become a major part of the cabinet, such as abrasives, adhesives, rags, and wax.

_____ 15. ____ of abrasives include sheets, belts, disks, sleeves, pads, or powders.
  A. Grits
  B. Types
  C. Forms
  D. None of the above.

_____ 16. Check an adhesives ____ before buying large quantities.

_____ 17. Buying tools is most easily done by ____.
  A. ordering by mail
  B. ordering by phone
  C. shopping at a store
  D. None of the above.

Name _____

Material and Supplies Order Form

Companies often design their own business forms. A material order form is a typical form that a company would design. In this activity, you will make a material order form. Create your form on this page. Fill in the form with the supplies and materials listed in Figure 21-1. Leave space at the bottom for totals. Figure costs using your own estimates of what you think the cost of various items would be.

Name _____ Date _____ Course _____

CHAPTER 22
Sawing with Hand and Portable Power Tools

Instructions: *Carefully read Chapter 22 of the text and answer the following questions.*

1. Name the two groups of handsaws.

   _____

_____ 2. Handsaws should be used to cut ____.
   A. high-density fiberboard
   B. particleboard
   C. plywood
   D. All of the above.

3. List four handsaws typically used for sawing straight lines.

   _____

   _____

   _____

   _____

_____ 4. A(n) ____ saw rips on one edge and crosscuts on the other.

5. What process is being done in the following photo?

Patrick A. Molzahn

_____

6. Identify the saw in the following drawing.

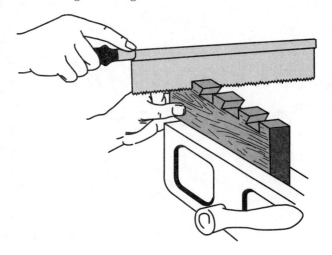

Porter

_____

7. Identify the saw in the following photo.

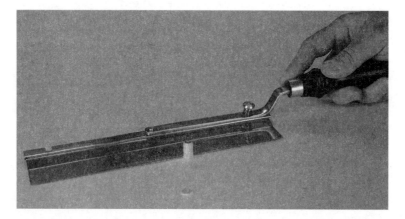

Patrick A. Molzahn

_____

8. What is the difference between a compass and a keyhole saw?
_____
_____
_____

Name _____

9. What type of saw is shown in the following photo?

Jeff Banke/Shutterstock.com

_____

_____ 10. *True or False?* Select a narrow blade saw for straight workpieces and a wide blade saw for curved workpieces.

_____ 11. Use a(n) ____ motion when sawing by hand.

12. List the steps in the hand sawing procedure.

_____
_____
_____
_____
_____
_____
_____
_____

13. Name four ways to support material before and after it is cut with a portable power saw.

_____
_____
_____
_____

_____ 14. The ____ saw is excellent for cutting lumber and paneling to approximate size.

15. Name two adjustments on the portable circular saw.

_____

_____

_____ 16. To prevent tearout when cutting paneling or plywood with a circular saw, ____ the stock with a razor knife on the side of the blade opposite the offcut.

_____ 17. Track saws have ____ on both sides of the blade to ensure that the cut has virtually no tearout.

_____ 18. ____ cuts are made to cut slots or pockets in the material.

_____ 19. ____ saws are frequently used to cut stiles and rails to length.
   A. Combination
   B. Power miter
   C. Track
   D. Reciprocating

20. Identify the saw in the photo below.

Bosch Power Tools

_____

_____ 21. When sawing curved lines, make ____ in the offcut along the curve.

Name _____

22. Identify the saw in the photo below.

Patrick A. Molzahn

_____

_____ 23. Proper maintenance of saw blades includes keeping them free of ____.
  A. moisture
  B. rust
  C. resin
  D. All of the above.

24. List four defects to check for when inspecting saw blades.

_____

_____

_____

_____

Name _____

Which Tool to Use

Study the side view of the desk in the following drawing, and then respond to the following statements.

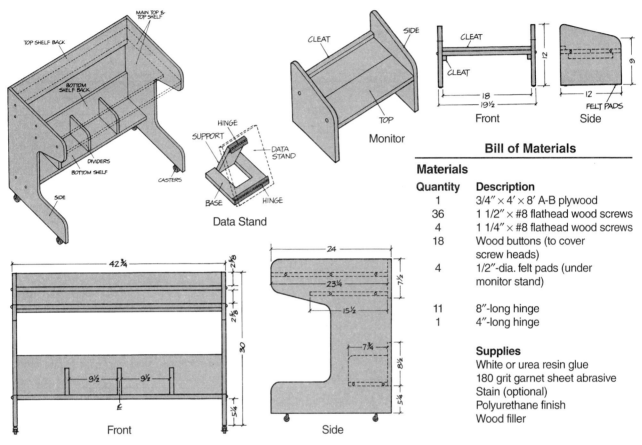

APA—The Engineered Wood Assoc.

If you had to cut this part using hand tools, explain what tools you would use and the procedure you would follow.

_____

_____

_____

_____

If you had to cut this part using portable power tools, explain what tools you would use and the procedure you would follow.

_____

_____

_____

_____

Name _____ Date _____ Course _____

CHAPTER 23
Sawing with Stationary Power Machines

Instructions: *Carefully read Chapter 23 of the text and answer the following questions.*

1. Stationary power-sawing machines are designed for what type of cuts.
   _____
   _____

2. Name four items to consider when selecting the proper saw.
   _____
   _____
   _____
   _____

3. List six safety tips to follow for safe and efficient operation of a power saw.
   _____
   _____
   _____
   _____
   _____
   _____

_____ 4. The most accurate straight-line sawing is done on equipment having a(n) ____ blade.

_____ 5. When sawing straight lines, the material must be ____ before and after the cut.

6. Identify the parts of the table saw.

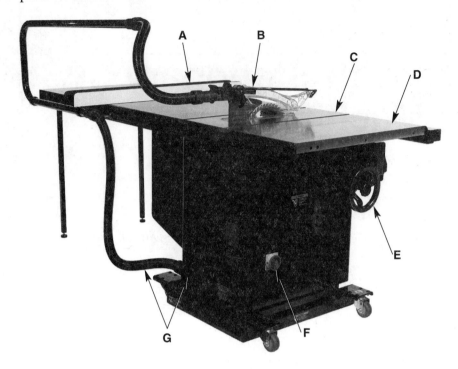

SawStop LLC

A. _____
B. _____
C. _____
D. _____
E. _____
F. _____
G. _____

7. Name the four major components of a tilting arbor table saw.

_____
_____
_____
_____

_____ 8. A miter gauge guides material at angles other than _____ to the blade.

9. Why is a blade guard essential?

_____
_____

_____ 10. *True or False?* A rip blade saws across the grain.

Name _____

_____ 11. Use a(n) ____ when the saw cut is shorter than the length of the material.
   A. miter gauge
   B. rip fence
   C. splitter
   D. overhead guide

_____ 12. When ripping plywood using a carbide-tipped rip blade, set the blade height ____.
   A. 1″ (25 mm) above the panel thickness
   B. even with the panel thickness
   C. 1/4″–1/2″ (6 mm–13 mm) above the panel thickness
   D. 3/4″ (19 mm) above the panel thickness

_____ 13. Sawing with the blade tilted is known as ____.

_____ 14. *True or False?* The workpiece dimensions on a bevel-cut edge will be different on the top and bottom faces.

_____ 15. ____ creates two or more thin pieces from thicker wood on edge.

_____ 16. Grooves cut perpendicular to the grain are known as a(n) ____.
   A. ploughs
   B. dados
   C. relief cuts
   D. gullets

_____ 17. To cut full-size sheets, most cabinetmakers use a(n) ____ rather than a table saw.

_____ 18. The ____ saw is most noted for sawing stock to length.
   A. band
   B. radial arm
   C. scroll
   D. tilting-arbor table

19. When installing a blade and tightening the arbor nut on a saw, why should you avoid overtorquing the nut?

_____

_____

_____ 20. The versatility of the radial arm saw comes from its wide range of ____.

_____ 21. The radial arm saw motor and blade assembly tilts 45° left and right for ____.

156   Modern Cabinetmaking Lab Workbook

_____ 22. Choose a(n) ____ saw for cutting large radius curves and large cabinet components.
   A. radial arm
   B. band
   C. scroll
   D. tilting-arbor table

_____ 23. Both band saw and scroll saw operations require relief cuts for making curves when ____.
   A. there is a sharp inside or outside curve
   B. the curve changes direction
   C. the cabinet part will be cut from a large piece of stock
   D. All of the above.

24. Identify the parts of the band saw.
   A. _____
   B. _____
   C. _____
   D. _____
   E. _____
   F. _____
   G. _____
   H. _____

Delta International Machinery Corp.

_____ 25. On a band saw, ____ position and control the blade above and below the table.

26. What is a U-shaped cut?

_____

_____

Name _____

27. Identify the parts of the scroll saw.

A. _____
B. _____
C. _____
D. _____
E. _____

Delta International Machinery Corp.

28. Why should you select a scroll saw blade that will have three or more teeth in contact with the wood at all times?

_____

_____

29. Name two types of cuts made with a scroll saw.

_____

_____

30. Why is the scroll saw the only stationary saw capable of easily cutting interior openings?

_____

_____

31. Why should you inspect the cut edges of a workpiece after making a cut?

_____

_____

_____

32. Chip load on a saw blade depends on four factors. Name those factors.

_____

_____

_____

_____

33. List four specifications for circular blades.

_____

_____

_____

_____

34. Identify the tooth and blade shapes.

A. _____

B. _____

C. _____

D. _____

E. _____

F. _____

G. _____

Carbide Tipped Blades

Tooth Shape

A   B   C

Blade Shape

D   E

F   G

_____ 35. The ____ is where chips accumulate as teeth cut through the material.
    A. dado
    B. hook angle
    C. grind
    D. gullet

_____ 36. A(n) ____ saw blade is an endless bonded loop of thin steel with teeth on one edge.
    A. radial arm
    B. scroll
    C. band
    D. circular

Name _____

37. Identify the band saw blade sets and shapes.

A. _____
B. _____
C. _____
D. _____
E. _____
F. _____

38. Identify the scroll saw blades.

A. _____
B. _____
C. _____
D. _____

_____ 39. The standard blade length for a scroll saw is ____.

_____ 40. ____ grinding is the only method to sharpen a carbide-tipped blade.

_____ 41. When using a table saw, an out-of-square workpiece might indicate a(n) ____ problem.
  A. table
  B. miter gauge
  C. fence
  D. All of the above.

_____ 42. Accurate adjustments can be made on a table saw using a(n) ____.

_____ 43. Which of the following can be a source of maintenance problems on radial arm saws?
  A. Rust
  B. Lack of lubrication
  C. Excessive torque on levers
  D. All of the above.

44. Name four adjustments that can be made on a band saw.

_____

_____

_____

_____

_____ 45. ____ saw blades can be coiled for storage.

_____ 46. Check the airflow periodically on ____ saws, because the pump could be damaged or the hose could be loose or broken.

Name _____

Using Stationary Power Tools

You are to make a cabinet part that is a finished size of 3/4″ × 16″ × 20″ from a full sheet of 4′ × 8′ plywood. Cut 6″ radius corners on all four corners. Number and explain the steps you would take to make this part with stationary power tools.

Name _____ Date _____ Course _____

CHAPTER **24**
Surfacing with Hand and Portable Power Tools

Instructions: *Carefully read Chapter 24 of the text and answer the following questions.*

_____ 1. *True or False?* You must read the wood grain after planing.

_____ 2. Always scrape away excess adhesive from joints ____.
   A. before planing
   B. during planing
   C. after planing
   D. None of the above.

3. Identify the parts of the bench plane shown below.

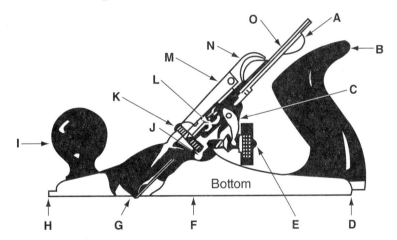

Stanley Tools

A. _____   I. _____
B. _____   J. _____
C. _____   K. _____
D. _____   L. _____
E. _____   M. _____
F. _____   N. _____
G. _____   O. _____
H. _____

4. Name four types of bench planes.

_____

_____

_____

_____

_____ 5. *True or False?* The jack plane has less tendency to follow the contours of warped lumber.

_____ 6. Bench plane adjustments control the ____, which is the cutter.

_____ 7. When planing a surface, you want ____.
   A. thick chips
   B. thin and feathery shavings
   C. sawdust
   D. None of the above.

_____ 8. When planing a surface, use ____ strokes with the grain.

9. Which type of plane work well for flattening cupped boards?

_____

_____ 10. When planing an edge, ____.
   A. mark a line to which you will cut
   B. plane against the grain figure
   C. hold the plane square
   D. All of the above.

_____ 11. Compared to bench planes, a(n) ____ plane is shorter, lighter, and has fewer moving parts.

12. Identify the parts of the block plane.

greseil/Shutterstock.com

A. _____   D. _____
B. _____   E. _____
C. _____   F. _____

Name _____

_____ 13. The plane iron on a block plane is set at a lower angle than it is on a bench plane.

_____ 14. Scrapers are effective on ____.
   A. edge grain
   B. end grain
   C. face grain
   D. Both A and C.

_____ 15. *True or False?* When scraping, the tool is behind the cutting edge.

16. Identify each scraper edge shown.

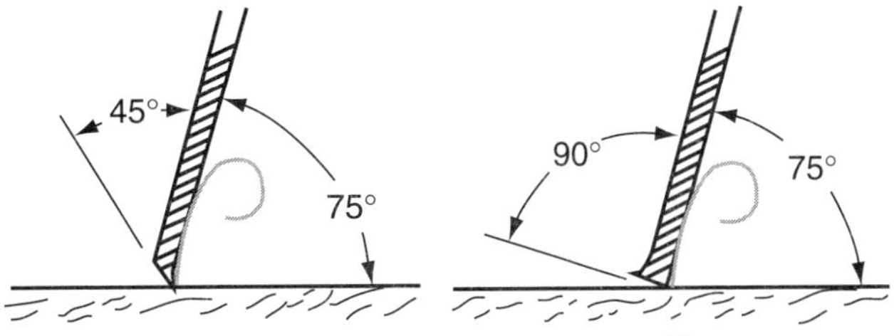

   A   B

A. _____

B. _____

17. List four safety tips to follow when surfacing with hand tools.

_____

_____

_____

_____

18. Identify the parts of the portable power plane.

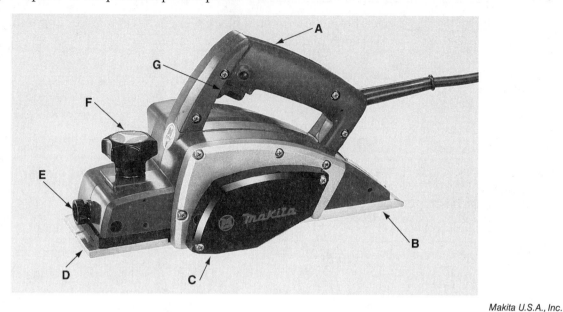

A. _____

B. _____

C. _____

D. _____

E. _____

F. _____

G. _____

19. What two adjustments are necessary when setting up the portable power plane?
_____
_____

_____ 20. When using the portable power plane, the last cut made on a surface should be ____ or less.
   A. 1/4"
   B. 1/8"
   C. 1/16"
   D. 1/32"

21. Why is it important to remove rust from metal surfaces?
_____

Name _____

The Hand Plane

In the space below, explain why you agree or oppose the following statement: The hand plane has no place in the modern production cabinetmaking shop.

_____
_____
_____
_____
_____
_____
_____
_____
_____
_____
_____
_____
_____
_____
_____
_____
_____
_____
_____
_____
_____
_____
_____
_____
_____

Name _____ Date _____ Course _____

CHAPTER 25
Surfacing with Stationary Machines

Instructions: *Carefully read Chapter 25 of the text and answer the following questions.*

_____ 1. Surfacing usually corrects ____ wood.

_____ 2. *True or False?* In wood with cross-grain, the lines formed by the annual rings run parallel to each other the full length of the board.

3. Look at the diagram shown. Indicate what type of grain it is and in what direction it is fed during surfacing.

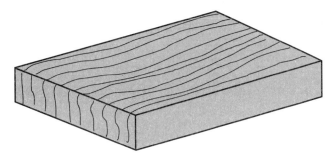

Grain: _____

Direction of feed: _____

_____ 4. The jointer is a multipurpose tool for surfacing ____.
  A. face
  B. edge
  C. end grain
  D. All of the above.

5. Identify the parts of the jointer.

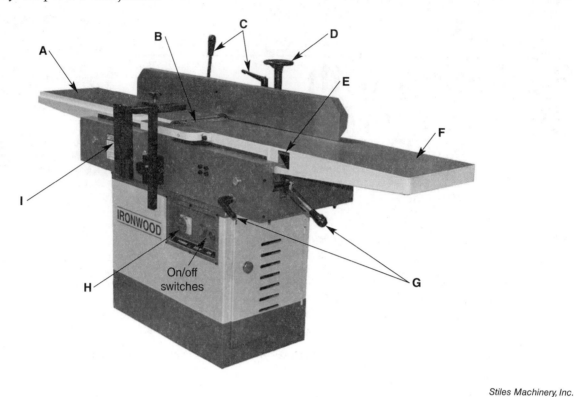

A. _____
B. _____
C. _____
D. _____
E. _____
F. _____
G. _____
H. _____
I. _____

_____ 6. When preparing for a jointer surfacing, set the jointer fence at ____, using a square.
  A. 15°
  B. 45°
  C. 60°
  D. 90°

7. When jointing, what is the minimum thickness the material should be? Why?
_____
_____

_____ 8. *True or False?* When jointing a cupped workpiece, place the cupped side down, to prevent the material from rocking.

Name _____

_____ 9. When jointing an edge, check to see that the fence is at a(n) ____ angle to the table.

_____ 10. When jointing ____, advance the end about 1" (25 mm) into the cutterhead, then lift and turn the workpiece around.

11. What operation is being shown here?

Chuck Davis Cabinets

_____

_____ 12. To make the second face of a board parallel to the first, use a(n) ____.

Match each term with the correct description

_____ 13. Reduce friction between the workpiece and the table.
_____ 14. Hold stock against the table after the cut is made.
_____ 15. Hold the workpiece down and reduces splitting.
_____ 16. Additional material removed from the leading or trailing end of a board.
_____ 17. Grabs the stock to pull it out from under the cutterhead.

A. Pressure bar
B. Infeed roller
C. Table rollers
D. Chip breaker
E. Outfeed roller
F. Snipe

172   Modern Cabinetmaking Lab Workbook

18. Identify the parts of the planer.

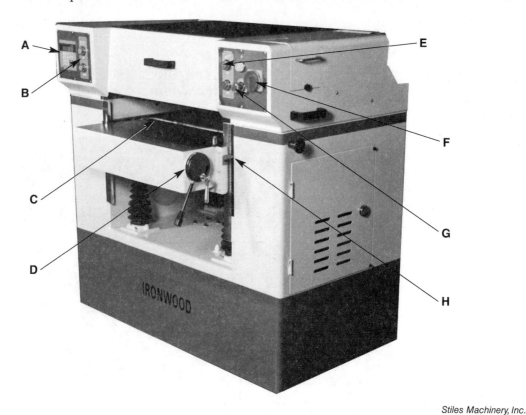

Stiles Machinery, Inc.

A. _____
B. _____
C. _____
D. _____
E. _____
F. _____
G. _____
H. _____

19. List three steps to prepare a planer for surfacing.

_____

_____

_____

_____ 20. *True or False?* For hardwoods, such as oak, the feed rate should be faster than for softwoods, such as pine.

_____ 21. When planing glued stock, ____.
    A. use a feed rate slower than normal
    B. use a normal feed rate
    C. use a feed rate faster than normal
    D. None of the above.

Name _____

_____ 22. The minimum thickness for planing stock without a backing board is ____.
    A. 1″ (25 mm)
    B. 3/8″ (9.5 mm)
    C. 1/4″ (6 mm)
    D. 1/8″ (3 mm)

23. Identify the machine shown here.

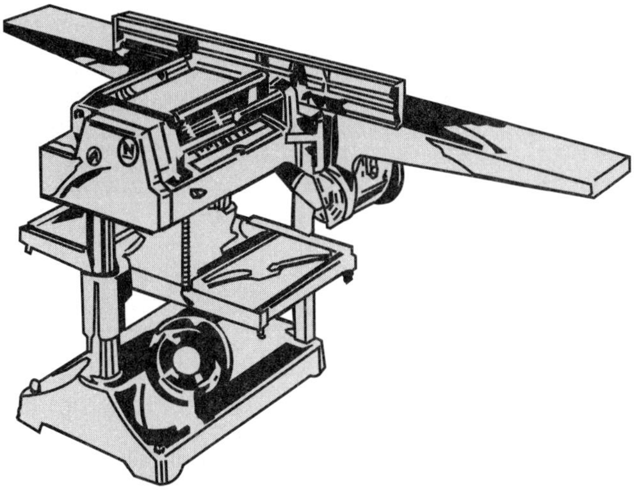

_____

_____ 24. Surfacing machines should be kept ____.
    A. clean
    B. properly adjusted
    C. lubricated
    D. All of the above.

25. Before surfacing, help prevent damage to cutting edges by checking for what four items?

_____

_____

_____

_____

_____ 26. *True or False?* The sharpness of planer and jointer knives can be determined by listening to the machine.

_____ 27. Sparks that fly ____ the wheel indicate a dull edge.

28. Identify the following jointer operation problems.

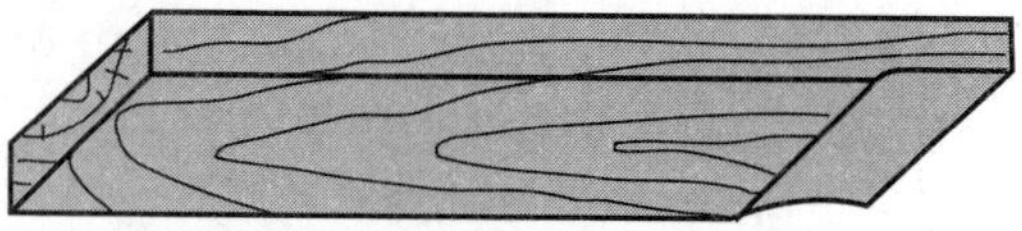

   A. _____
   B. _____ 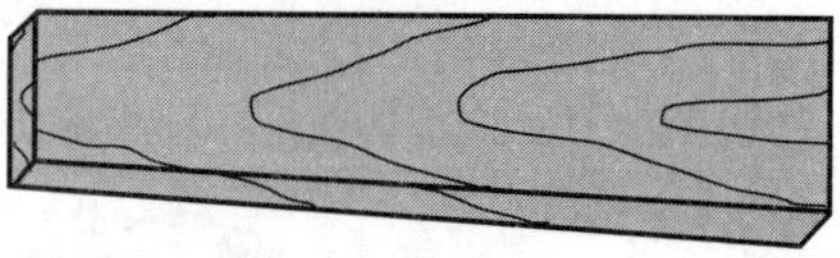
   C. _____ 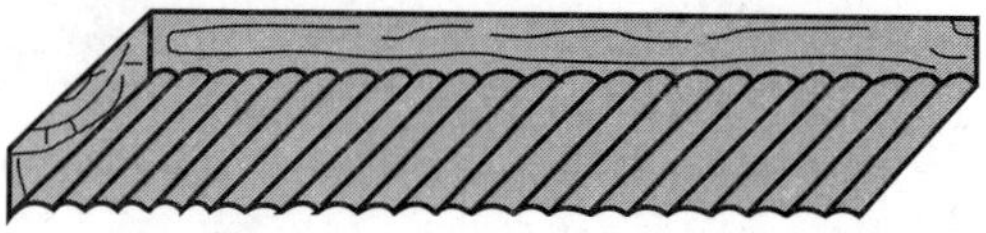
   D. _____ 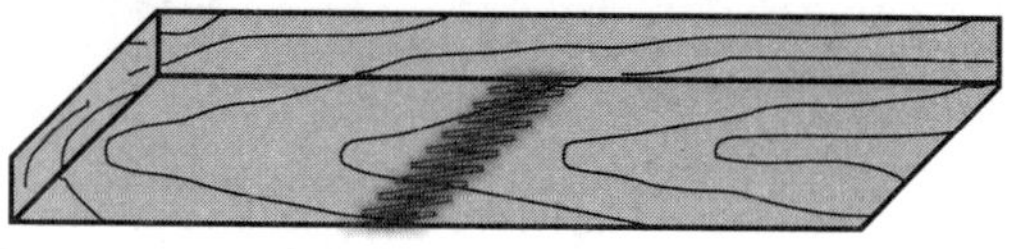

_____ 29. ____ restores the cutting edge to planer knives.

30. List six adjustments that must be checked after planer knives are sharpened.

_____
_____
_____
_____
_____
_____

Name _____

Troubleshooting and Correcting Planer Problems

Read the problems and causes in the planer troubleshooting chart shown here. Complete the chart by identifying at least one solution to each problem.

Troubleshooting Hints		
Problem	**Cause**	**Solution**
Board will not feed through.	1. Pressure bar too low. (most common cause) 2. Table rollers too low. 3. Insufficient pressure on infeed roller or outfeed roller. 4. Cut too deep.	
Snipe appears at beginning of board only.	1. Front table roller set too high.	
Snipe appears at end of board only.	1. Rear table roller set too high.	
Chip appears 3–6″ (7.6–15.2 cm) from both ends of the board.	1. Pressure bar set too high. 2. Table roller set too high.	
Board appears to splinter out.	1. Excessive feed. 2. Cutting against grain. 3. Chipbreaker too high. 4. Green lumber.	
Knives raise grain.	1. Dull knives. 2. Green lumber.	
Chip marks appear on stock.	1. Exhaust system not working properly. 2. Loose connection in exhaust system. 3. Chips stuck on outfeed roller.	
Taper across width.	1. Table not parallel with cutterhead.	

(Continued)

Troubleshooting Hints		
Problem	**Cause**	**Solution**
Glossy or glazed surface appearance on stock.	1. Dull knives. 2. Too slow a feed.	
Washboard surface finish.	1. Knives not set at the same height. 2. Too fast a feed rate. 3. Table gibs loose.	
Chatter marks across width of board. (Small washboard.)	1. Table rollers too high (particularly) noticeable on thin material.	
Line on workpiece parallel to feed direction.	1. Nick in knives. 2. Scratch in pressure bar.	
Excessive noise.	1. Dull knives. 2. Joint on knives too wide. 3. Table roller too high for workpiece thickness.	
Excessive vibration.	1. Knives not sharpened evenly such that they are different heights.	
Workpiece twists while feeding.	1. Pressure bar not parallel. 2. Table rollers not parallel with table. 3. Uneven pressure on infeed or outfeed roller. 4. Chipbreaker not parallel. 5. Resin buildup on table.	
Main drive motor kicks out.	1. Excessive cut. 2. Bad motor. 3. Dull knives.	
Feed motor stalls.	1. Bad motor. 2. Lack of lubrication on idlers.	

Powermatic

Name _____ Date _____ Course _____

CHAPTER **26**
Shaping

Instructions: *Carefully read Chapter 26 of the text and answer the following questions.*

_____ 1. Which of the following tools are designed strictly for shaping?
   A. Shapers
   B. Moulders
   C. Routers
   D. All of the above.

2. Identify the parts of the spindle shaper shown here.
   A. _____
   B. _____
   C. _____
   D. _____
   E. _____
   F. _____
   G. _____
   H. _____
   I. _____

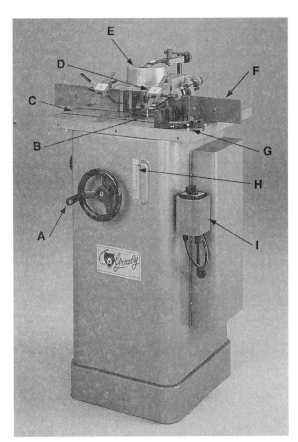

Grizzly Imports, Inc.

_____ 3. When hand feeding material into a spindle shaper, it must always be fed against the ____ rotation.

4. Identify each cutter shaper.

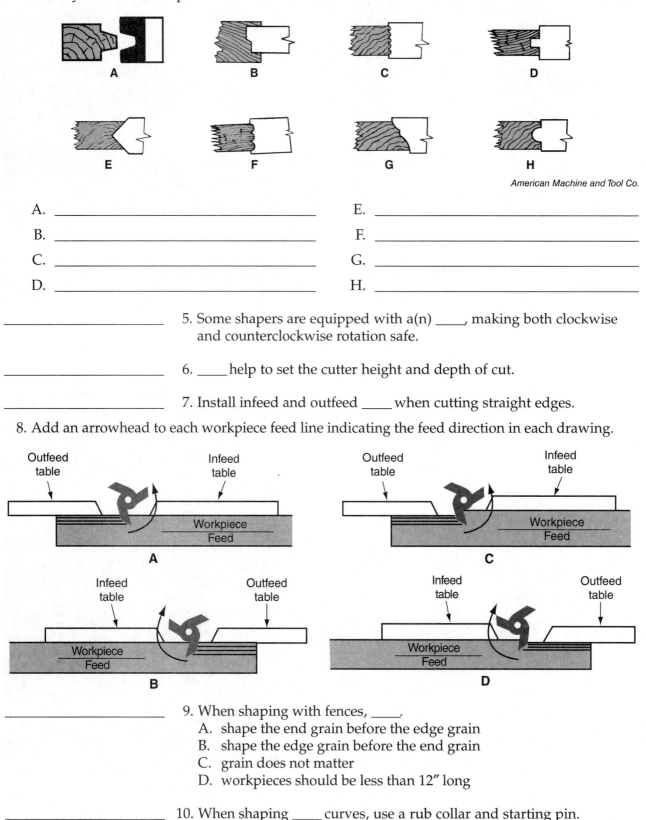

American Machine and Tool Co.

A. _____   E. _____
B. _____   F. _____
C. _____   G. _____
D. _____   H. _____

_____ 5. Some shapers are equipped with a(n) ____, making both clockwise and counterclockwise rotation safe.

_____ 6. ____ help to set the cutter height and depth of cut.

_____ 7. Install infeed and outfeed ____ when cutting straight edges.

8. Add an arrowhead to each workpiece feed line indicating the feed direction in each drawing.

_____ 9. When shaping with fences, ____.
   A. shape the end grain before the edge grain
   B. shape the edge grain before the end grain
   C. grain does not matter
   D. workpieces should be less than 12″ long

_____ 10. When shaping ____ curves, use a rub collar and starting pin.

_____ 11. ____ are patterns used to duplicate workpieces.

Name _____

_____ 12. The angle jig supports the workpiece at an angle for shaping ____.
A. bevels
B. miters
C. Both A and B.
D. None of the above.

_____ 13. Cove cutting is a shaping operation done on a(n) ____.

14. Identify the parts of the router bit shown here.

A. _____
B. _____
C. _____

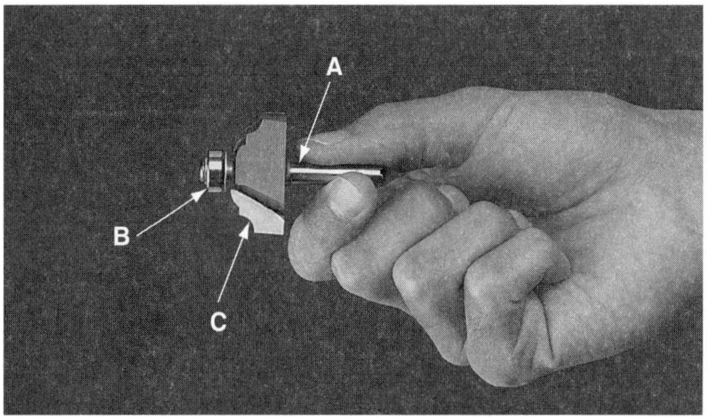

Bosch

_____ 15. Lowering the router bit into the workpiece is called ____.

16. Identify the typical router bit shapes and the shapes they create by writing the name of the shape next to each one.

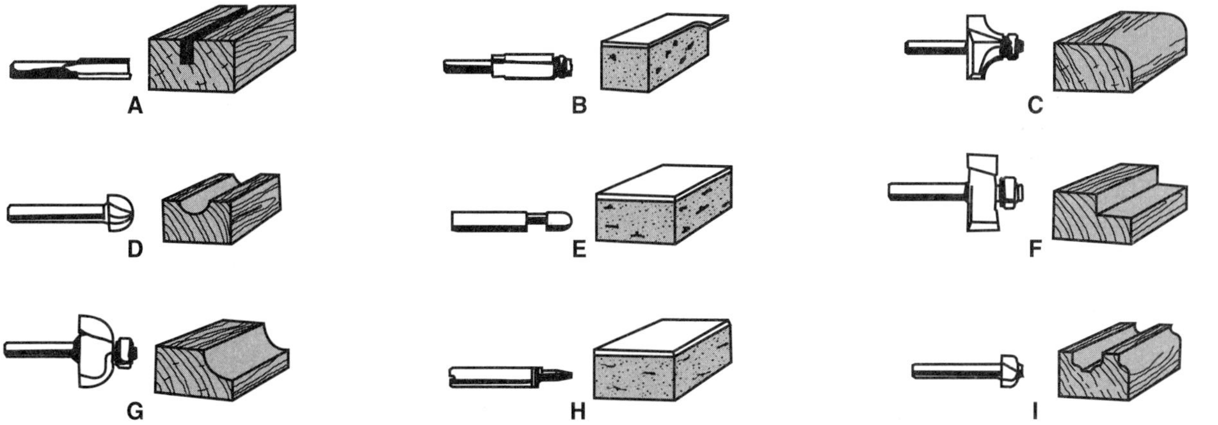

Bosch

A. _____
B. _____
C. _____
D. _____
E. _____
F. _____
G. _____
H. _____
I. _____

180 Modern Cabinetmaking Lab Workbook

_____ 17. The ____ router is much like an overarm router, except that the bit is located below the workpiece.

18. Identify the parts of the inverted router.

   A. _____
   B. _____
   C. _____
   D. _____
   E. _____
   F. _____

C. R. Onsrud

_____ 19. When installing router bits, engage the ____ so the motor shaft will not turn as you tighten the collet.

20. Which part guides the bit in the following diagram?

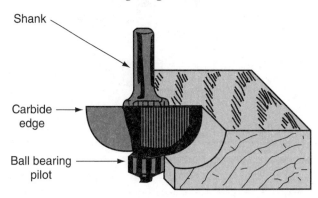

Chuck Davis Cabinets

_____

_____ 21. To rout a dado, clamp a(n) ____ to the workpiece.

Copyright Goodheart-Willcox Co., Inc.
May not be reproduced or posted to a publicly accessible website.

Name _____

_____ 22. Which of the following is used to shape intricate areas?
　　　　　　　　　A. Laminate trimmer
　　　　　　　　　B. Motorized rotary tools
　　　　　　　　　C. Spokeshave
　　　　　　　　　D. None of the above.

23. Identify this tool.

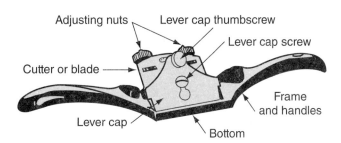

Stanley Tools

_____

24. Identify this tool.

Patrick A. Molzahn

_____

_____ 25. ____ have perforated metal blades with many individual cutting edges.

_____ 26. Removing built-up resins on cutters with ____ will prolong the life of the cutting edge.

_____ 27. Grind cutters only when the cutting lip is ____.

Name _____

Shaping an Edge

Imagine you have just made a tabletop for a coffee table and you would like to add a profile on the edge. Draw a cross section of one edge of the tabletop, with the edge added. Then explain the procedure you would follow to put the edge on the tabletop.

Cross section:

Procedure:

_____

_____

_____

_____

_____

_____

_____

Name _____ Date _____ Course _____

CHAPTER **27**
Drilling and Boring

Instructions: *Carefully read Chapter 27 of the text and answer the following questions.*

_____ 1. Drilling and boring are ____ processes.

2. Explain the difference between drilling and boring.

_____

_____

_____

_____

_____ 3. Which of the following drills and bits are the most cost-effective for cabinetmaking?
　　A. Carbide-tipped
　　B. High-speed steel
　　C. Carbon steel
　　D. They are all equal.

_____ 4. Drilling is done with ____.
　　A. hand tools
　　B. portable power drills
　　C. stationary machines
　　D. All of the above.

_____ 5. Auger bits are *not* effective for boring holes in ____ grain.

6. Identify the parts of the auger bit shown here.

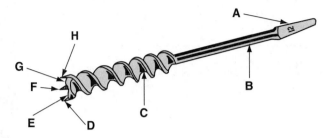

A. _____
B. _____
C. _____
D. _____
E. _____
F. _____
G. _____
H. _____

_____ 7. *True or False?* Auger bits are fast speed bits.

8. What type of bit is shown here?

Irwin

_____

9. Identify the parts of the twist drill.

Greenlee

A. _____
B. _____
C. _____

_____ 10. ____ bits drill a flat-bottomed hole.
    A. Machine spur
    B. Brad point
    C. Spade
    D. Multispur

Name _____

Match each term with the correct description.

_____ 11. Effective for drilling pilot holes for screws or nails.

_____ 12. Used to drill extra-long holes.

_____ 13. Make plugs to cover mechanical fasteners in counterbored holes.

_____ 14. Designed for boring holes in wood for pipe and conduit.

_____ 15. Have carbide tips to drill holes in concrete and ceramic materials.

_____ 16. Angles hole tops to allow flat head screws to sit flush with the surface of the stock.

_____ 17. Creates large holes through workpieces.

_____ 18. Is flat and has a fairly long brad point center to it.

_____ 19. Creates a flatter hole bottom than a machine spur bit creates.

_____ 20. Includes combination drills and countersink/counterbore cutters.

_____ 21. Twist bit surrounded by a spring-loaded retractable sleeve.

_____ 22. Makes holes in concrete for cabinet installation.

_____ 23. Has a guide drill and the circumference is a series of saw teeth.

_____ 24. Allow user to make long holes with standard drills and bits.

A. Circle cutter
B. Star drill
C. Drill extensions
D. Multi-operational bits
E. Brad point bit
F. Spade bit
G. Hole saw
H. Plug cutter
I. Countersink bit
J. Vix bit
K. Multispur bit
L. 32mm System boring bit
M. Drill points
N. Bell hanger's drill
O. Masonry drill
P. Glass drill

25. Identify the parts indicated on the brace to the right.

A. _____
B. _____
C. _____
D. _____
E. _____
F. _____
G. _____
H. _____
I. _____
J. _____
K. _____

Stanley Tools

_____ 26. A(n) ____ is operated with one hand.
A. brace
B. hand drill
C. push drill
D. drill press

_____ 27. The ____ is the most common vertical stationary power drill used by cabinetmakers.

28. What type of stationary power machine is designed for the manufacture of European-style cabinetry?
_____

29. Name two types of portable power tools for drilling.
_____
_____

30. What is the purpose of covering vise jaws with wood blocks when drilling and boring?
_____
_____

_____ 31. *True or False?* A blind hole goes through the workpiece.

32. Identify the drill attachments shown here.

Patrick A. Molzahn

A. _____
B. _____

Name _____

33. Explain how to bore a hole at an angle.

_____

_____

_____

34. Put the following steps for drilling and boring with the drill press in order.

_____ A. Clamp workpiece to table when drilling large holes, holes at an angle, or when using saw tooth and adjustable bits.

_____ B. Insert the bit in the chuck.

_____ C. Unlock the quill, slide stock to the side of the bit, and adjust the depth stop.

_____ D. Adjust the table.

_____ E. Set the drill press speed.

_____ F. Lower the bit with the feed lever and align the tip with the hole layout marks. Lock the quill.

_____ G. Adjust the guard to within 1/4″ (6 mm) of workpiece.

_____ H. Turn motor on and lower bit into workpiece.

_____ I. Raise bit and turn off machine.

_____ 35. Drill holes at an angle by ____.
   A. tilting the machine table
   B. using a fixture
   C. using a jig
   D. Both A and B.

36. Explain two methods for drilling equally spaced holes on the drill press.

_____

_____

_____

_____

_____

_____

_____ 37. Before sharpening, check to see that the bit is ____.

_____ 38. The preferred method for sharpening an auger bit is with an auger bit ____.

_____ 39. *True or False?* Twist drills are difficult to sharpen by hand.

_____ 40. Spade bits can be sharpened by ____.
  A. grinding
  B. honing
  C. filing
  D. All of the above.

_____ 41. Portable drills need grease in the ____ gears.

Name _____

Cabinet Installer

Imagine you are a cabinet installer. Explain in detail what drills and bits you would need in your toolbox and why.

_____
_____
_____
_____
_____
_____
_____
_____
_____
_____
_____
_____
_____
_____
_____
_____
_____
_____
_____
_____
_____
_____
_____
_____
_____
_____
_____
_____
_____
_____

Name _____ Date _____ Course _____

CHAPTER **28**
Computer Numerically Controlled (CNC) Machinery

Instructions: *Carefully read Chapter 28 of the text and answer the following questions.*

1. What does *CNC* stand for?

   _____

_____ 2. ____ is used to move materials and products throughout a facility and for repetitive processes like sanding.

_____ 3. CNC machinery is driven by ____.

4. What is shown in the following image?

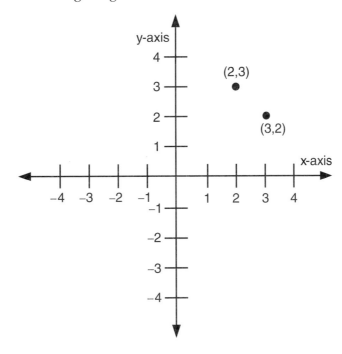

_____ 5. Any tool that can be equipped with a(n) ____ for adjusting settings is capable of being controlled by a computer.

6. Name two types of drive mechanisms that are common on CNC machines.

_____

_____

_____ 7. *True or False?* Nested base routers require adjustments for varying material sizes.

_____ 8. ____ machining centers require that parts be cut to approximate size before machining.
   A. Flat table
   B. Pod-and-rail
   C. Vertical
   D. Cellular

_____ 9. The ____ is the brain behind the CNC machine.

10. List four safety barriers often found on CNC machines that reduce the chance of injury to the user.

_____

_____

_____

_____

Match each step in the CNC process with the item that performs that step.

_____ 11. Visual representation of the product (geometry)

_____ 12. Holds the part and executes the operations

_____ 13. Delivers motion instructions to the motors which control the various axes

_____ 14. Converts geometry into meaningful machinery instructions with tool info

_____ 15. Translates machining instructions into G-code which can be read by a specific machine

A. CAD
B. CAM
C. Post-processor
D. CNC controller
E. CNC machine

_____ 16. The tool ____ is the location of the cutting tool relative to the geometry.

_____ 17. ____ is a universally accepted set of motion commands used by many CNC machine tools.
   A. G-code
   B. Conversational programming
   C. Drawing interchange format
   D. Parametric programming

Name _____

_____ 18. A type of generic programming allows the programmer to use computer-related features to create programs that can be reused for multiple part sizes.
   A. G-code
   B. Conversational programming
   C. Drawing interchange format
   D. Parametric programming

19. Explain the function of a spoilboard in the CNC machining process.

_____

_____

_____

_____

20. What is onion-skinning?

_____

_____

_____

_____

_____ 21. The speed at which the tool moves through the material is known as _____.

22. What is chip load? How does it affect tool life?

_____

_____

_____

_____

23. Calculate the spindle speed in rpm that would need to be programmed for a CNC router to cut a 3/4″ (19.05 mm) thick MDF pattern using a feed rate of 860 IPM (inches per minute) and 1/2″ double-fluted bit in the spindle. Round answer to the nearest 1000. Note: use the upper limit chip size from the chart in the book.

_____

_____

_____

_____

196 Modern Cabinetmaking Lab Workbook

_____ 24. *True or False?* Carbide is the most common material used for CNC tooling.

_____ 25. ____ are devices that store tools for quick retrieval during machining.

26. Name two measurements the machine control must know in order to calculate the correct cutting depth and offset.

_____

_____

27. What type of manufacturing is shown in the following image?

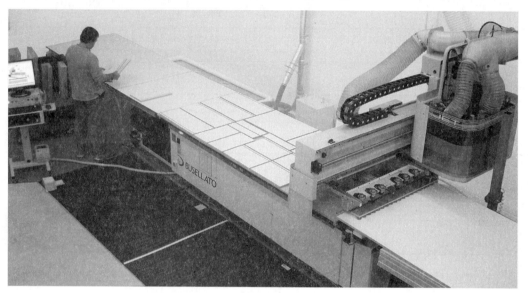

Casadei Busellato

_____

_____ 28. The beam saw, shown here, is typical of the machines used in ____ manufacturing.

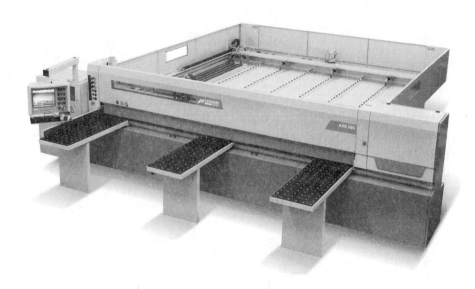

Casadei Busellato

Copyright Goodheart-Willcox Co., Inc.
May not be reproduced or posted to a publicly accessible website.

Name _____

Figuring CNC Coordinates

List the coordinates of the points that the center point of a 1/2" diameter CNC router bit would travel to in order to cut the part shown here. Use a straight tool path (refer to Figure 28-28). The Z axis positioning should not be considered in this activity. Point A is Point 0, 0. Start your cutting at Point A, and proceed clockwise around the piece.

Refer to Figure 28-5 for explanation of the Cartesian coordinate system.

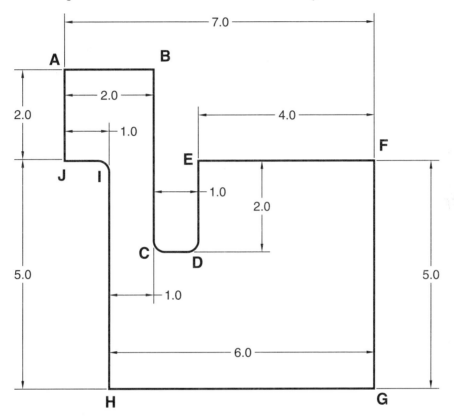

FILLETS ARE .25 RADIUS

Point A: X −0.25   Y 0.25
Point B: X   2.25   Y 0.25
Point C: X   2.25   Y −3.75
Point D: X   2.75   Y −3.75
Point E: X   2.75   Y −1.75
Point F: X   7.25   Y −1.75
Point G: X   7.25   Y −7.25
Point H: X   0.75   Y −7.25
Point A: X −0.25   Y 0.25

Name _____ Date _____ Course _____

CHAPTER **29**
Abrasives

Instructions: *Carefully read Chapter 29 of the text and answer the following questions.*

_____ 1. Abrasives are used by cabinetmakers to ____ surfaces in preparation for assembly or finishing.

_____ 2. ____ abrasives are grains of a natural mineral or synthetic substance bonded to a cloth or paper backing.

_____ 3. ____ abrasives are grains bonded into stones and grinding wheels.

_____ 4. ____ abrasives, such as pumice and rottenstone, are finely crushed abrasives.

5. Explain how natural abrasives differ from synthetic abrasives.

_____

_____

_____

_____

_____

6. Natural abrasives used by cabinetmakers are listed below. Describe them.

Garnet _____

_____

_____

_____

Emery _____

_____

_____

_____

Pumice _____
_____
_____
_____

Rottenstone _____
_____
_____
_____

Tripoli _____
_____
_____
_____

7. Some of the synthetic abrasives used by cabinetmakers are listed below. Describe them.

Aluminum oxide _____
_____
_____
_____

Silicon carbide _____
_____
_____
_____

Industrial diamond _____
_____
_____
_____

_____ 8. ____ size refers to the screen holes per linear inch.

_____ 9. Which of the following is the heaviest paper backing?
    A. A
    B. C
    C. D
    D. E

Name _____

10. Complete the following chart about abrasive grains.

CAMI Grit No.	FEPA Grit No.	Common Name	Application
600 500 400 360 320 280 240 220	P1200 P1000 P800 P600 P500 P400 P360 P320 P280 P240 P220		
180 150 120	P180 P150 P120		
100 80 60	P100 P80 P60		
50 40 36	P50 P40 P36		
30 24 20 16	P30 P24 P20		

Klingspor Abrasives, Inc.

_____ 11. ____-grade cloth backings are for belt and hand sanding machines.

_____ 12. ____-grade cloth backings are used for flat sanding belts on large production machinery.

_____ 13. Heavy-duty disk and drum abrasives are made with ____ backing.

_____ 14. Abrasive grains are attached to the backing between two separate layers of ____.

_____ 15. On a closed coat abrasive, ____ of the backing is covered with abrasive grains.
   A. the entire surface
   B. 50%
   C. 70%
   D. 80%

_____ 16. When bonded by ____, grains of 150 grit or coarser dropped directly on a wet make coating.

17. Identify each type of flexing shown here.
   A. _____
   B. _____
   C. _____
   D. _____

Match each abrasive form with the correct description.

_____ 18. Scored sheet of abrasive bound to a center core with a shaft.   A. Disk
_____ 19. Available in diameters of 4 1/2" through 12" (114 mm through 305 mm). B. Sleeve
                                                                                  C. Belt
_____ 20. Coated abrasive bonded to a flexible sponge.   D. Block
_____ 21. Made for arbor-mounted drum and spindle sanders.   E. Flap wheel
_____ 22. Range in size from less than 1" (25 mm) to 52" (1.32 m) wide.   F. Pencil
_____ 23. Used for sanding channels, recesses, and bottoms of blind holes.   G. Spiral
                                                                                  H. Roll

_____ 24. ____ abrasives are a combination of abrasives and bonding agents.

_____ 25. Grinding wheels are most commonly a vitrified bond of ____ oxide grains.

_____ 26. Sharpening and honing tools often are made with natural and synthetic ____.

Name _____ Date _____ Course _____

CHAPTER 30
Using Abrasives and Sanding Machines

Instructions: *Carefully read Chapter 30 of the text and answer the following questions.*

_____ 1. Smoothing with ____ is the process of removing surface wood cells to achieve a smooth, blemish-free surface.

_____ 2. Which grain requires the most abrading?
   A. Face grain
   B. Edge grain
   C. End grain
   D. None of the above.

_____ 3. Look for ____ caused by planing or scraping before abrading.

4. How are abrasives selected?

_____

_____

_____

_____

_____ 5. *True or False?* When selecting the proper grit size, each heavier grit size removes the marks of the abrasive before it.

_____ 6. Increase grit size by no more than ____ grade numbers at a time below 150.

_____ 7. Sand ____ the grain with the final grit to help hide abrasive scratches.

8. Explain what is being done in the photo below.

_____

_____

9. Name two ways to sand contours.

_____

_____

_____ 10. _____ sanders use an abrasive belt that travels around a drive roller and on one or more idler rollers.
   A. Disk
   B. Drum
   C. Dual-action
   D. Edge

11. Identify the parts of the disk sander shown here.

   A. _____
   B. _____
   C. _____
   D. _____

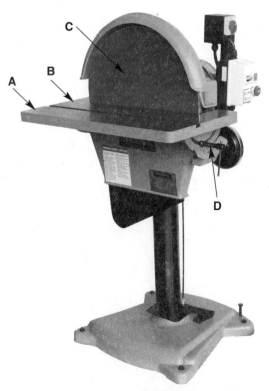

General International Mfg. Ltd.

Name _____

_____ 12. *True or False?* When using the disk sander, only use the half of the disk that rotates upward.

13. What type of sanding machine is shown in the following image?

Courtesy of Tadsen Photography for Madison College

_____

_____ 14. ____ sanders are designed to flatten panels.
   A. Portable profile
   B. In-line finishing
   C. Drum
   D. Spindle

15. List five machine- or abrasive-related defects caused by wide belt sanding.

_____

_____

_____

_____

_____

Match each sanding tool with the correct description.

_____ 16. Consists of abrasive strips or flat pieces of abrasive mounted on a wheel.

_____ 17. Also known as a dual-action sander.

_____ 18. Electric motor turns a pair of drums on which an abrasive belt is mounted.

_____ 19. Moves the abrasive back and forth in a straight line.

_____ 20. Provides in-line motion to remove machining marks on concave, convex, or flat profiles.

_____ 21. Practical where grain direction is not a factor.

A. Portable belt sander
B. Portable finishing sander
C. In-line finishing sander
D. Orbital-action finishing sander
E. Random orbital finishing sander
F. Portable profile sander
G. Flap sander
H. Portable drum sander

22. Name two tools that can be used to remove dust from all moving parts, electrical boxes, and motor vents on sanding equipment.

_____

_____

_____ 23. Do not leave _____ disks on sanders after use because they can leave a residue that will cause irregular pressure when sanding.

Name _____

Setting Up Shop with Abrasive Machines

Imagine you are setting up a small, two-person cabinetmaking shop. You are responsible for choosing which abrasives and sanding machines you will have in the shop. Choose three machines you must have. Explain how they will be used in your shop.

Machine 1 _____

_____
_____
_____
_____
_____
_____
_____

Machine 2 _____

_____
_____
_____
_____
_____
_____
_____

Machine 3 _____

_____
_____
_____
_____
_____
_____
_____

Name _____ Date _____ Course _____

CHAPTER **31**
Adhesives

Instructions: *Carefully read Chapter 31 of the text and answer the following questions.*

_____ 1. Which of the following is an adhesive?
   A. Cement or glue
   B. Mastic
   C. Resin
   D. All of the above.

_____ 2. ____ occurs by adding a liquid substance that wets the surfaces to be joined and then hardens to withstand stress on the assembly.

3. Identify each bonding process.
   A. _____
   B. _____

_____ 4. Select adhesives according to the materials they will ____.

5. What is shelf life?
_____
_____

_____ 6. _____ describes the amount of time you have to fit workpieces together after applying adhesive.
   A. Assembly time
   B. Clamp time
   C. Curing time
   D. Shelf life

_____ 7. The time after the adhesive sets until the joint reaches full strength is known as _____.

_____ 8. *True or False?* Metal and plastics are porous materials.

9. What is a product data sheet?

_____

_____

_____ 10. The _____ less surface contact between workpieces, the better the joint.

_____ 11. Which of the following affects the durability of a joint?
   A. Temperature
   B. Moisture
   C. Stress
   D. All of the above.

12. Name three forms of wood adhesives.

_____

_____

_____

_____ 13. Ease of application is known as _____.
   A. wet tack
   B. sandability
   C. spreadability
   D. gap filling ability

Name _____

14. Read the following chart and then identify each type of glue being described.

Comparison of Typical Ready-Use Adhesives			
Characteristic	A	B	C
Appearance	Cream	Clear white	Clear amber
Spreadability	Good	Good	Fair
Acidity (pH level*)	4.5–5.0	4.5–5.0	7.0
Speed of Set	Very fast	Fast	Slow
Stress Resistance†	Good	Fair	Good
Moisture Resistance	Fair	Fair	Poor
Heat Resistance	Good	Poor	Excellent
Solvent Resistance‡	Good	Poor	Good
Gap Filling Ability	Fair	Fair	Fair
Wet Tack	High	None	High
Working Temperature	45°F–110°F	60°F–90°F	70°F–90°F
Film Clarity	Translucent	Very clear	Clear but amber
Film Flexibility	Moderate	Flexible	Brittle
Sandability	Good	Fair (will soften)	Excellent
Storage (shelf life)	Excellent	Excellent	Good

pH—glues with a pH of less than 6 are considered acidic and could stain acidic woods such as cedar, walnut, oak, cherry, and mahogany.
† Stress resistance—refers to the tendency of a product to give way under constant pressure.
‡ Solvent resistance—ability of finishing materials such as varnishes, lacquers, and stains to take over a glued joint.

Franklin International

A. _____

B. _____

C. _____

_____ 15. Casein and plastic resin are the two basic types of ____ adhesives.

16. What is a thermoset bond?

_____

_____

_____ 17. Two-part adhesives use a(n) ____ to harden a resin, forming an adhesive compound.

18. Read the following chart and then identify each type of glue being described.

Comparison of Typical Water-Mixed and Two-Part Adhesives			
Characteristic	A	B	C
Appearance	Cream	Tan	Clear to Amber
Spreadability	Fair	Excellent	Good
Speed of Set	Slow	Slow	†Slow to Fast
Stress Resistance	Good	Good	Excellent
Moisture Resistance	Good	Good	Waterproof
Heat Resistance	Good	Good	Excellent
Solvent Resistance	Good	Good	Excellent
Gap Filling Ability	Fair to Good	Fair	Excellent (nonshrinking)
Wet Tack	Poor	Poor	Poor to Fair
Working Temperature	32°–110°F	70°–100°F	50°F–120°F
Film Clarity	Opaque	Opaque	†Clear to Amber
Film Flexibility	Tough	Brittle	Tough
Sandability	Good	Good	Good
Storage (shelf life)	1 year	1 year	Unlimited (if unmixed)

† Depending on formulation

Franklin International

A. _____

B. _____

C. _____

_____ 19. Contact cement works well as an adhesive on all of the following except ____.
   A. wood
   B. the decorative face of plastic laminate
   C. metal
   D. ceramic

Match each type of adhesive with the correct description.

_____ 20. A form of thermoplastic adhesive, it is designed to be melted in a hot-glue gun.

_____ 21. Used primarily for bonding plastic laminate to particleboard or MDF substrate.

_____ 22. Bond unfinished and prefinished plywood, hardboard, and similar panels to wood, metal and concrete.

_____ 23. Also known as super or instant glues.

_____ 24. Nontoxic and nonflammable, will not damage lacquered, painted, or varnished surfaces.

A. Solvent-based contact cement
B. Chlorinated-based contact cement
C. Water-based contact cement
D. Panel adhesive
E. Vinyl-based adhesive
F. Cyanoacrylate adhesive (CA)
G. Hot-melt adhesive

Name _____

_____ 25. Fitting an assembly together before applying glue is known as a(n) ____ run.

26. What is being done in the photograph below?

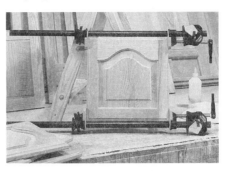

American Tool Companies

_____

_____

_____ 27. In which of the following situations has glue been applied properly?
A. Small beads of glue ooze from the joint.
B. No glue oozes from the assembly.
C. Glue drips from the joint.
D. None of the above.

_____ 28. Glue from an electric gun cools and sets in approximately ____ seconds.
A. 15
B. 30
C. 45
D. 60

29. What process is taking place in the following photo?

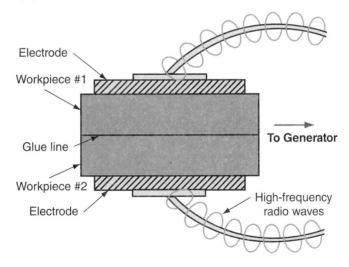

_____

_____ 30. Which of the following adhesives is *not* adaptable for RF gluing?
   A. Urea formaldehyde resin
   B. Cross-linking polyvinyl acetate resin
   C. Hide glues
   D. Aliphatic resin

Name _____

Supply the Adhesive

Imagine you are setting up a small, two-person cabinetmaking shop. You are responsible for ordering essential adhesives for the shop. What three adhesives would you choose and why?

Adhesive 1 _____

_____

_____

_____

_____

_____

_____

_____

Adhesive 2 _____

_____

_____

_____

_____

_____

_____

_____

Adhesive 3 _____

_____

_____

_____

_____

_____

_____

_____

_____

Name _____ Date _____ Course _____

CHAPTER 32
Gluing and Clamping

Instructions: *Carefully read Chapter 32 of the text and answer the following questions.*

1. List four ways cabinetmakers can use clamps.

   _____

   _____

   _____

   _____

2. Identify the type of clamp shown below.

   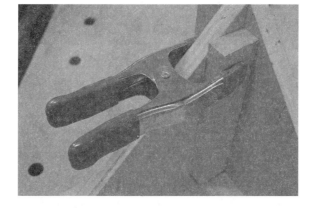

   Patrick A. Molzahn

   _____

3. Name a disadvantage of using screw clamps.

   _____

   _____

   _____

4. Identify the type of clamps shown below.

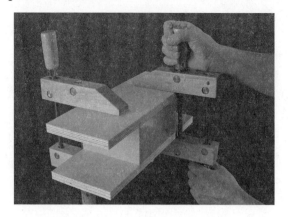

Patrick A. Molzahn

_____

5. Identify the type of clamps shown below.

Patrick A. Molzahn

_____

6. Identify the type of clamps shown below.

Bessey Tools North America

_____

Name _____

7. Identify the type of clamp shown below.

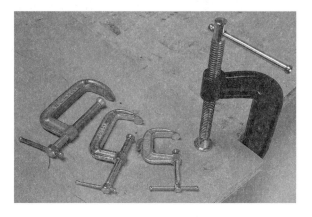

Patrick A. Molzahn

_____

8. Identify the type of clamp shown below.

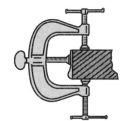

Shopsmith

_____

_____ 9. Which of the following clamps are essential for irregular shapes?
    A. C clamps
    B. Web clamps
    C. Band clamps
    D. Both B and C.

10. Identify the type of clamp shown below.

Bessey Tools North America

_____

11. Identify the type of clamp shown below.

Patrick A. Molzahn

_____

12. Identify the type of clamp shown below.

Bessey Tools North America

_____

13. Identify the type of clamp shown below.

Patrick A. Molzahn

_____

Name _____

14. Identify the type of clamp shown below.

Patrick A. Molzahn

_____

15. Identify the type of clamp shown below.

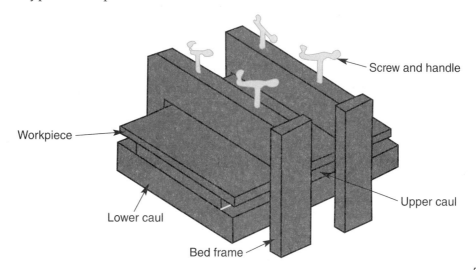

Franklin International

_____

16. Identify the type of clamp shown below.

American Tool Companies

_____

17. Identify the type of clamp shown below.

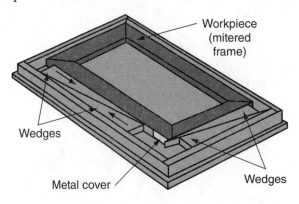

_____

_____ 18. Backer blocks and ____ protect your product from damage caused by clamp jaws.

19. Explain why you should cut and place wax paper between clamps or backer blocks and the workpiece at a glue joint.

_____

_____

_____

20. List the four primary steps in the clamping procedure.

_____

_____

_____

_____

21. When edge-gluing workpieces, why should you alternate the end grain with each piece?

_____

_____

_____ 22. When clamping workpieces face-to-face, the most common problem with hand screw clamps is that the jaws are not aligned ____ to the approximate opening of the assembly.

_____ 23. When clamping frames, it is important that you work on a(n) ____ surface.

Name _____ Date _____ Course _____

CHAPTER 33
Bending and Laminating

Instructions: *Carefully read Chapter 33 of the text and answer the following questions.*

_____ 1. When choosing wood for bending, select wood with ____.
   A. straight grain
   B. knots
   C. figured grain
   D. splits

_____ 2. *True or False?* Moistening increases bendability of wood.

_____ 3. ____ is specialized plywood available for radius work.

4. List three items that make wood easier to bend.

_____

_____

_____

_____ 5. Wood's attempt to return to its original shape is ____.

6. Name two ways to bend lumber dry.

_____

_____

_____ 7. Workpieces to be bent dry should pass a(n) ____.

_____ 8. *True or False?* A kerf-bent curve is strong.

9. Identify the types of curves below.

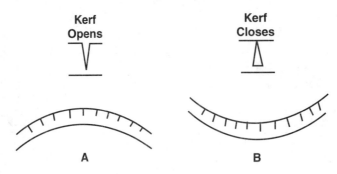

A. _____

B. _____

10. Name the three measurements you must find when kerf bending.

_____

_____

_____

_____ 11. _____ bending involves softening, or plasticizing, the lumber.

_____ 12. When wet bending, for good bendability, the final moisture content of wood should be ____.
   A. between 5% and 10%
   B. between 20% and 30%
   C. 100%
   D. None of the above.

13. Identify the types of wet bending molds below.

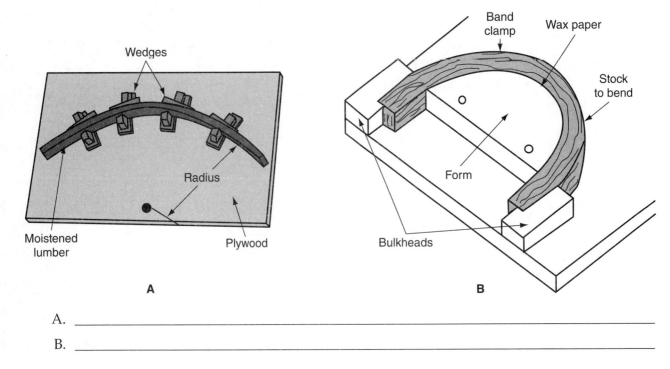

A. _____

B. _____

Name _____

_____ 14. When wet bending, pressure should be kept on bent wood ____.
   A. for one hour
   B. for one day
   C. until it dries
   D. None of the above.

_____ 15. Wood ____ is the process of bonding two or more layers of lumber or veneer.

16. List three reasons laminating is done.

_____
_____
_____
_____
_____
_____

17. Give four examples of common laminations.

_____
_____
_____
_____

_____ 18. *True or False?* Most curved product laminating is done with successive layers having different grain direction.

_____ 19. ____ are used primarily for structural laminations, such as beams and plywood.

20. Look at the illustration below. Which face is best for straight laminations?

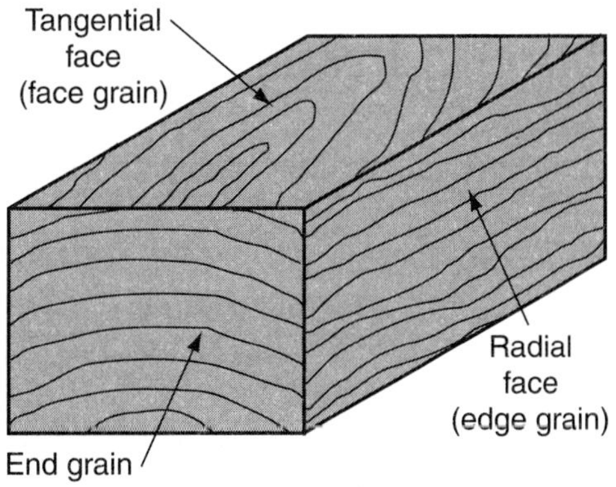

_____

_____ 21. *True or False?* Adhesives for laminating should have a long set time.

_____ 22. ____ laminations are done to increase strength and/or increase thickness.

23. List three ways of making curved laminations.

_____

_____

_____

_____ 24. With full surface, one direction laminations, ____ is not reversed so it is easier to bend.

_____ 25. ____ laminations are curves built of rows of solid wood pieces.

Name _____

Designing for Bending and Laminating

Design a small product out of wood that makes use of one of the processes discussed in the chapter. The design process should stop at the preliminary idea stage. Dimensions are not necessary. Explain in detail the bending or laminating process you would use in making the project.

Preliminary sketches:

Explanation of processes:

_____

_____

_____

_____

_____

_____

_____

_____

_____

_____

Name _____ Date _____ Course _____

CHAPTER **34**
Overlaying and Inlaying Veneer

Instructions: *Carefully read Chapter 34 of the text and answer the following questions.*

1. Name and describe two types of overlaying.

   _____

   _____

   _____

   _____

2. Name and describe two types of inlaying.

   _____

   _____

   _____

   _____

3. Identify the overlaying or inlaying processes illustrated below.

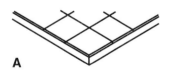  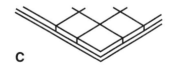 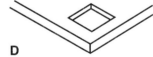

A   B   C   D

   A. _____

   B. _____

   C. _____

   D. _____

_____ 4. The surface to be decorated is called a(n) ____.

_____ 5. For inlaying, the substrate is normally ____.
  A. plywood
  B. surfaced solid wood workpieces
  C. fiberboard
  D. particleboard

_____ 6. Generally, rotary and flat slicing methods produce ____ grain in veneer.

7. How can you flatten veneer if it warps and distorts?

_____

_____

_____

8. Identify the tool below.

Patrick A. Molzahn

_____

_____ 9. When you need time to align veneer, which adhesive below is the best choice?
  A. Resin-based, water solvent adhesive
  B. Contact adhesive
  C. Hot melt glue
  D. None of the above.

_____ 10. Light, even ____ must be applied to an adhesive coated veneer to create a permanent bond.

_____ 11. Mostly, you will use ____ species and exotics for veneering.

Name _____

12. Look at the layouts below. Identify the matches that can result from them.

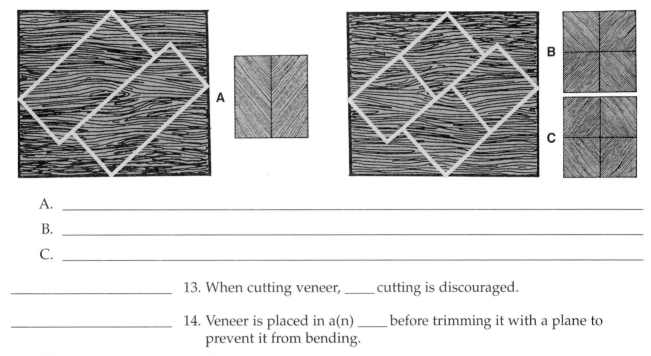

A. _____
B. _____
C. _____

_____ 13. When cutting veneer, _____ cutting is discouraged.

_____ 14. Veneer is placed in a(n) _____ before trimming it with a plane to prevent it from bending.

15. What is being done in the illustration below?

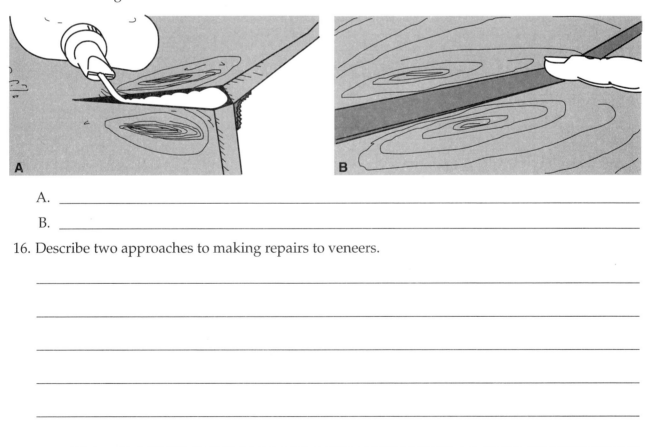

A. _____
B. _____

16. Describe two approaches to making repairs to veneers.

_____
_____
_____
_____
_____
_____

_____ 17. ____ is an overlaying process in which you cover the edges of manufactured panel products.

18. How does parquetry differ from veneering?

_____

_____

19. List the four steps involved in parqueting.

_____

_____

_____

_____

_____ 20. Most often, parquetry involves ____.
　　　　　　　　　　　　　　A. circles
　　　　　　　　　　　　　　B. curved designs
　　　　　　　　　　　　　　C. straight line geometric shapes
　　　　　　　　　　　　　　D. None of the above.

_____ 21. When cutting blocks for parquetry, accuracy is important and cuts are made with a(n) ____ tooth blade.

_____ 22. Use ____ adhesive when parqueting larger surfaces.

_____ 23. Most parqueted surfaces have a protective ____ surrounding the edge.

_____ 24. ____ are precut, decorative overlays.

25. List the six steps involved in the inlaying process.

_____

_____

_____

_____

_____

_____

26. Name two ways marquetry pieces can be cut.

_____

_____

Name _____

27. What is being done in the illustration below?

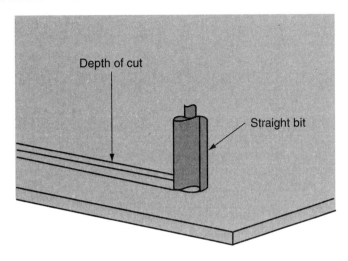

_____

28. Describe several special practices for finishing overlaid and inlaid surfaces that were discussed in the text.

_____
_____
_____
_____
_____
_____
_____

29. Describe three ways to prevent inlays from being discolored during finishing.

_____
_____
_____
_____
_____
_____

_____ 30. The production sequence for industrial veneering applications used in architectural millwork are similar to what is used in the small shop except the dimensions and _____ is much larger.

Name _____

Veneer Chess Board

Explain how you would make a chess board using veneers glued to a substrate material. Number your steps.

_____
_____
_____
_____
_____
_____
_____
_____
_____
_____
_____
_____
_____
_____
_____
_____
_____
_____
_____
_____
_____
_____
_____
_____
_____

Name _____ Date _____ Course _____

CHAPTER **35**

Installing Plastic Laminates

Instructions: *Carefully read Chapter 35 of the text and answer the following questions.*

1. Name two types of rigid plastic laminates.

   _____

   _____

   _____

_____ 2. Which of the following is *not* a good choice for a core material for plastic laminates?
   A Particleboard
   B Fiberboard
   C Solid lumber
   D Plywood

_____ 3. Most cabinet and countertop laminations are done with ____ (rigid, flexible) laminates because of their wear resistance.

_____ 4. Wipe the surface to be covered in laminate with ____ to clean it.

_____ 5. You can score and fracture rigid laminates as you would ____.

_____ 6. When cutting laminate, the saw teeth must enter the laminate on the ____ (decorative, back) side on the cutting stroke.

_____ 7. When two pieces of rigid laminate will be butted against each other, an accurate cut can be made with a(n) ____.

_____ 8. ____ is the primary adhesive for laminating.

_____ 9. When applying adhesive to laminates, the temperatures should be no
_____ lower than ____. The relative humidity should be between ____% and
_____ ____%.

_____ 10. Generally, you apply a laminate ____ before installing a(n) ____.

_____

_____ 11. ____ removes the sharp corners and extends the wear life of laminate edges.

_____ 12. A warm ____ or a(n) ____ can be used to heat laminate to make it flexible so that it can be laminated around a curve against a core.
_____

13. What is being done in the photo below?

Patrick A. Molzahn

_____

_____

Name _____

Laminating an Island Top

Write the procedure that you would use to laminate a kitchen bar counter with plastic laminate HPDL. The countertop has 1 1/2″ edge reinforcement and 4″ radius corners. Use numbered steps.

Name _____ Date _____ Course _____

CHAPTER **36**
Turning

Instructions: *Carefully read Chapter 36 of the text and answer the following questions.*

_____ 1. Turning processes produce round parts on a(n) ____.

2. Identify the two types of turning shown below.

Patrick A. Molzahn

A. _____

Patrick A. Molzahn

B. _____

3. Identify the parts indicated on the standard wood turning lathe below.

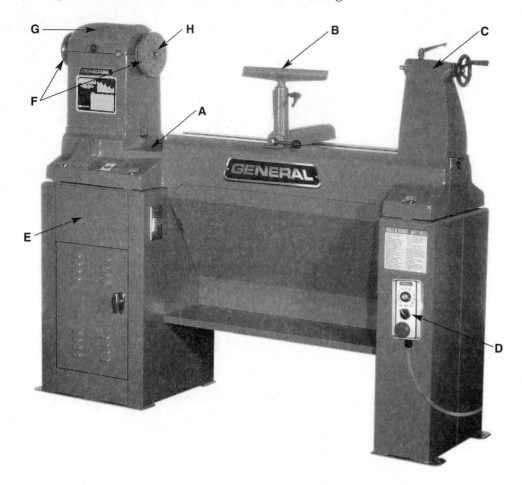

General International Mfg. Ltd.

A. _____
B. _____
C. _____
D. _____
E. _____
F. _____
G. _____
H. _____

Name _____

4. What type of lathe is shown below?

Teknatool U.S.A., Inc.

_____

5. Identify the turning tools shown below.

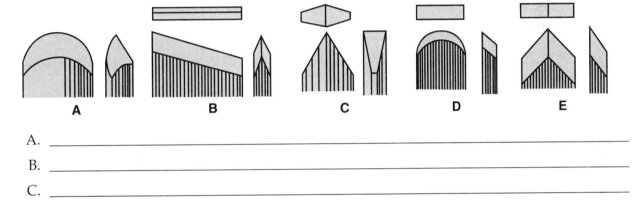

A. _____
B. _____
C. _____
D. _____
E. _____

6. Look at the illustration below. Which shows a cutting action and which shows a scraping action?

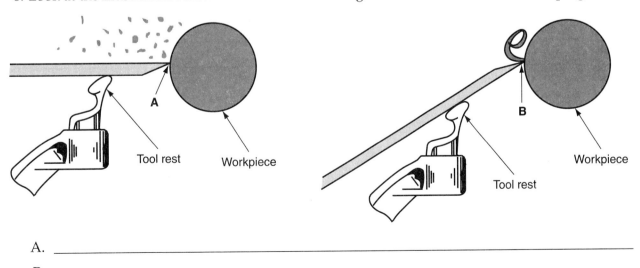

A. _____
B. _____

_____ 7. ____ direction affects the smoothness of a surface.

8. Identify the various accessories used for mounting stock.

A. _____
B. _____
C. _____
D. _____
E. _____
F. _____
G. _____

_____ 9. *True or False?* As a rule of thumb, use fast lathe speeds for large work and slow lathe speeds for small work.

_____ 10. Which of the following are the best choices for turning?
  A. Close grain hardwoods
  B. Open grain hardwoods
  C. Softwoods
  D. None of the above.

11. List at least five safety precautions to take when using a lathe.

_____
_____
_____
_____
_____

Name _____

12. Describe what is occurring in each of the illustrations.

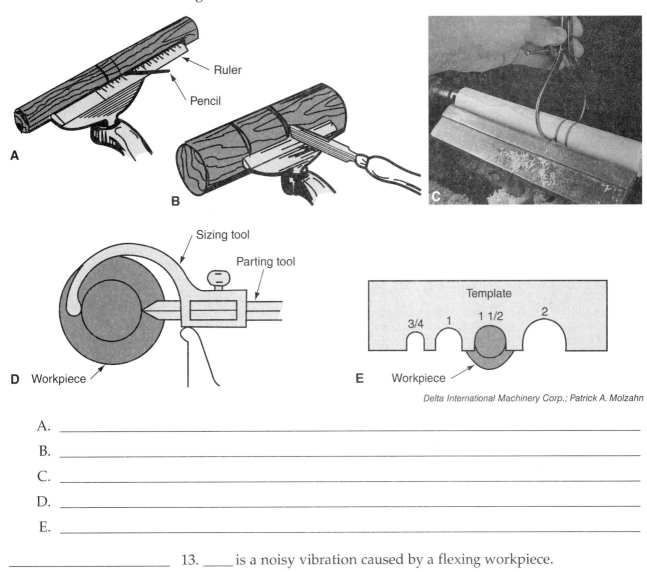

Delta International Machinery Corp.; Patrick A. Molzahn

A. _____
B. _____
C. _____
D. _____
E. _____

_____ 13. ____ is a noisy vibration caused by a flexing workpiece.

_____ 14. ____ are the same diameter from one end to the other.

15. What types of turning are taking place in the illustrations below?

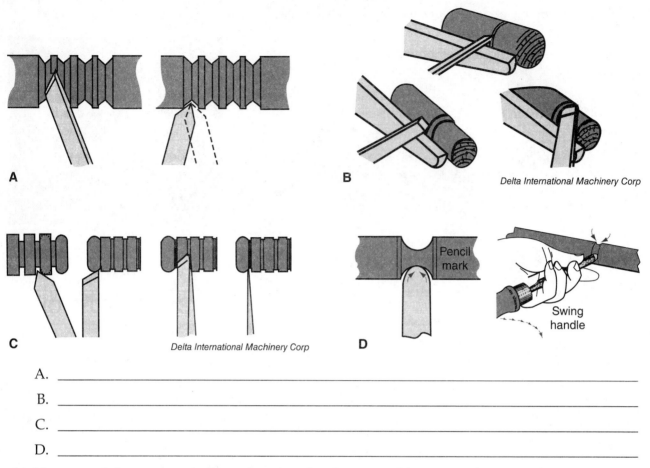

A. _____
B. _____
C. _____
D. _____

16. Next to each key pattern indicate the type of tool you would use in making this shape.

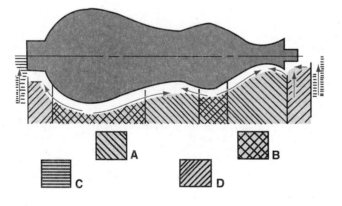

A. _____
B. _____
C. _____
D. _____

_____ 17. Turnings can be made of ____.
A. solid lumber
B. several layers of glued stock
C. Both A and B.
D. None of the above.

Name _____

_____ 18. ____ are glued-up workpieces that are separated after being turned.

_____ 19. If you are making spindles in a chair back, which of the following would be the most difficult duplicating practice?
   A. Using a template
   B. Measuring each detail independently
   C. Using a duplicator
   D. Both A and C.

_____ 20. You must turn oval spindles ____ times on ____ different centers.
_____

_____ 21. Which of the following are turned while attached to a faceplate?
   A. Bowls
   B. Stool seats
   C. Small round tabletops
   D. All of the above.

_____ 22. Material too large for inboard turning is mounted on the ____ side of the lathe.

_____ 23. Dowels and round workpieces are held easily in a lathe ____.

_____ 24. Frequent ____ keeps turning tools sharp and in good condition.

Match the following terms and descriptions for turning operations.

_____ 25. Moving the cutting edge of the turning tool toward the headstock.   A. In

_____ 26. Feeding the tool toward the centerline of the lathe.   B. Out

_____ 27. Moving the tool away from the centerline of the lathe mostly with the faceplate.   C. Left

_____ 28. Moving the cutting edge of the turning tool toward the tailstock.   D. Right

Name _____

Turning Projects

Brainstorm eight turning projects that can be made by turning wood on the lathe. Sketch your ideas below.

Name _____ Date _____ Course _____

CHAPTER **37**
Joinery

Instructions: *Carefully read Chapter 37 of the text and answer the following questions.*

_____ 1. One of the most important elements affecting the durability of a product is ____.

_____ 2. The direction of grain in a solid wood joint affects the ____ of the joint.

_____ 3. Which of the following joints is strongest?
  A. AG/AG
  B. AG/EG
  C. EG/EG
  D. They are all of equal strength.

_____ 4. Which of the following joints is the strongest?
  A. Radial grain bonded to radial grain
  B. Tangential grain bonded to tangential grain
  C. Radial grain bonded to tangential grain
  D. Both A and B are equal because they shrink at the same rates.

5. Name four joints that are suitable for manufactured panel products.

_____

_____

_____

_____

6. List the three types of joints.

_____

_____

_____

_____ 7. Which of the following is a non-positioned joint?
   A. Mortise and tenon joint
   B. Butt joint
   C. Dovetail joint
   D. None of the above.

8. Explain how positioned joints compare to non-positioned joints.

_____

_____

_____

_____ 9. ____ joints have some element in addition to adhesive that helps hold the joint.

_____ 10. It is best to choose the simplest joint that meets the ____ requirements of the product.

11. Identify the butt joints shown below.

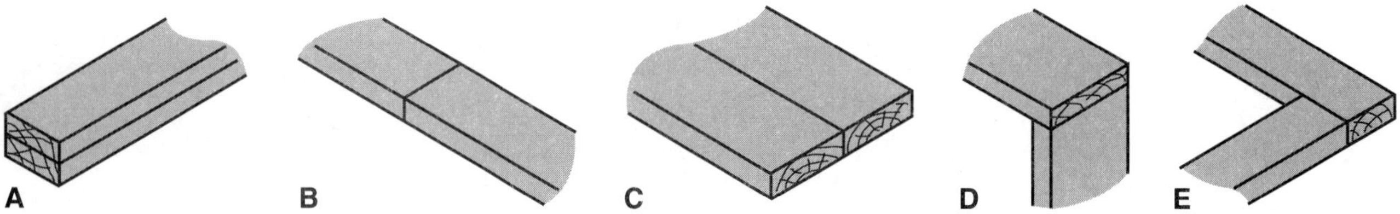

A. _____
B. _____
C. _____
D. _____
E. _____

_____ 12. *True or False?* Glue blocks make a butt joint weaker.

_____ 13. A(n) ____ joint is a slot cut across the grain.

_____ 14. A(n) ____ is a slot cut with the grain.

Name _____

15. Identify the dado joints below.

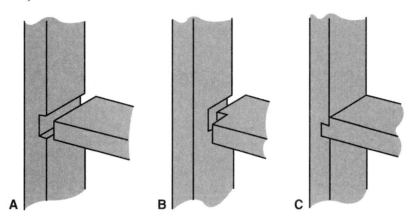

A. _____
B. _____
C. _____

16. Identify the type of joint shown below.

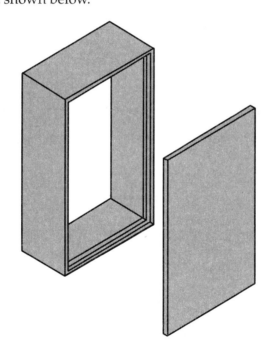

_____

17. Identify the types of joints below.

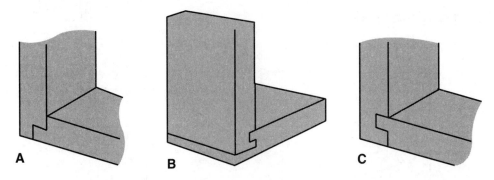

A. _____
B. _____
C. _____

18. Identify the types of lap joints below.

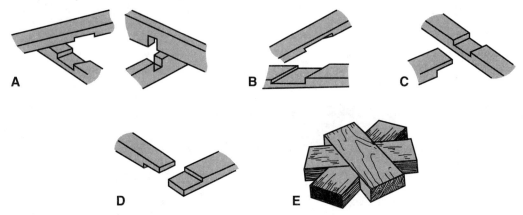

Delta International Machinery Corp.

A. _____
B. _____
C. _____
D. _____
E. _____

19. Identify the types of miter joints shown below.

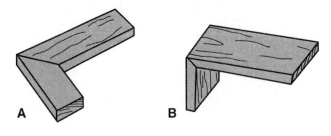

Delta International Machinery Corp.

A. _____
B. _____

Chapter 37 Joinery **255**

Name _____

20. Identify the type of joint shown below.

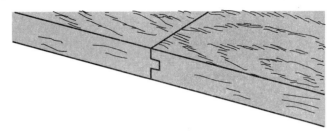

_____

21. Identify the following types of mortise and tenon joints.

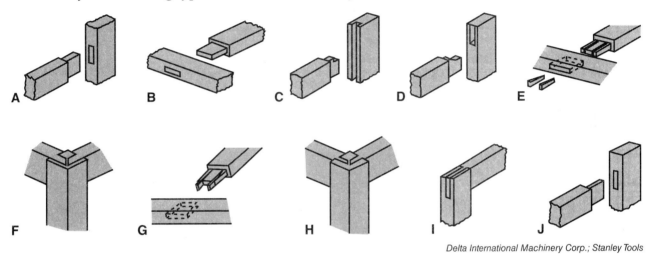

Delta International Machinery Corp.; Stanley Tools

A. _____

B. _____

C. _____

D. _____

E. _____

F. _____

G. _____

H. _____

I. _____

J. _____

22. Identify the decorative mortise and tenon joints below.

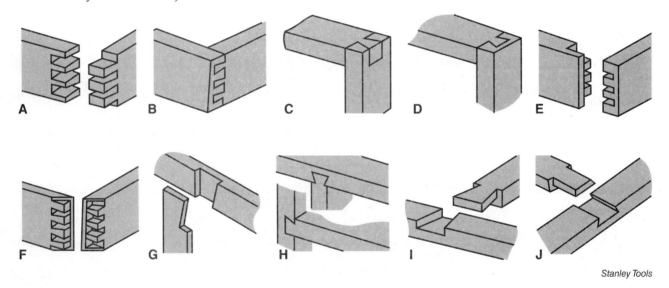

A. _____
B. _____
C. _____
D. _____

23. Identify the dovetail joint variations below.

A. _____
B. _____
C. _____
D. _____
E. _____
F. _____
G. _____
H. _____
I. _____
J. _____

Chapter 37 Joinery **257**

Name _____

24. Explain why dowel holes must be drilled accurately.

_____

_____

_____ 25. ____ joinery is excellent for connecting plastic-laminated materials that will not accept a glue bond.

26. Identify the various types of spline joints below.

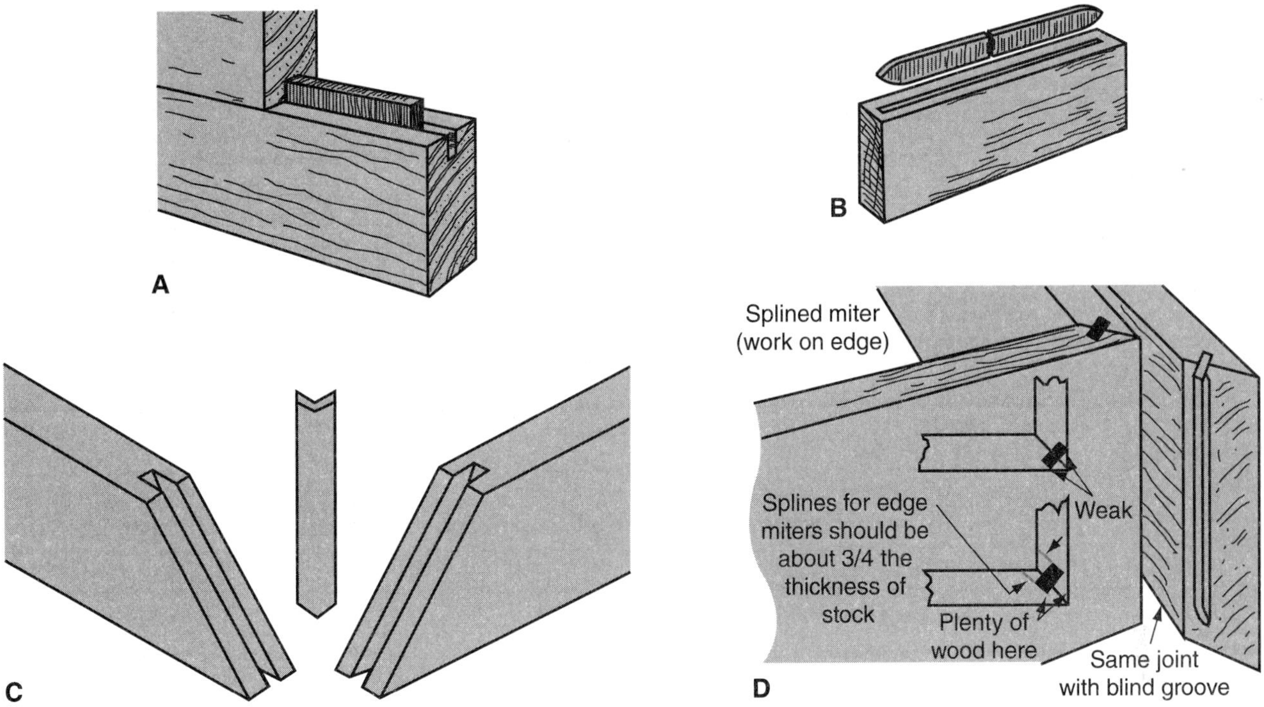

A. _____
B. _____
C. _____
D. _____

27. Identify the type of joint below.

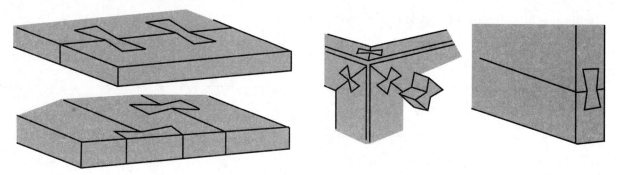

Shopsmith

_____

_____ 28. A(n) ____ joint allows the cabinetmaker to use screws to connect end-to-end or edge-to-edge.

29. Identify the joint below.

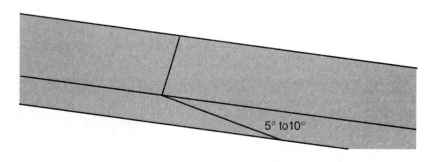

_____

30. Identify the joint below.

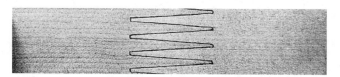

Forest Products Laboratory

_____

_____ 31. When threading, the ____ produces the external threads.

_____ 32. The ____ produces the internal threads when threading..

33. List four safety precautions to take when making joints.

_____

_____

_____

_____

Name _____

Choose a Joint

The following diagram is a plan view of a simple trinket box. What type of joint would you choose to fasten the parts together? Draw the joint in at each corner and explain why you chose that joint.

Explanation:

_____
_____
_____
_____
_____
_____
_____
_____
_____
_____
_____
_____

Name _____ Date _____ Course _____

CHAPTER **38**

Accessories, Jigs, and Special Machines

Instructions: *Carefully read Chapter 38 of the text and answer the following questions.*

_____ 1. Table attachments make saw operations ____.
   A. easier
   B. safer
   C. more accurate
   D. All of the above.

_____ 2. ____ help you support large stock.

_____ 3. A(n) ____ supports longer and heavier stock for cross and miter cuts.

_____ 4. ____ keep material against the table and fence and prevent kickback.

5. Identify the item below and briefly describe its purpose.

Patrick A. Molzahn

_____
_____
_____
_____

Copyright Goodheart-Willcox Co., Inc.
May not be reproduced or posted to a publicly accessible website.

6. Identify the item below and briefly describe its purpose.

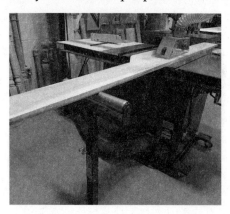

Patrick A. Molzahn

_____

_____

_____

_____ 7. A(n) ____ consists of a spindle adapter, two guides, and a ring base.

_____ 8. ____ hold a workpiece in position and also guide the tool or workpiece.

_____ 9. ____ are holding devices that do not guide the tool.

10. Identify the item below and briefly describe its purpose.

The Fine Tool Shops

_____

_____

_____

Name _____

11. Identify the item below and briefly describe its purpose.

Patrick A. Molzahn

_____
_____
_____

_____ 12. ____ allow you to cut wedges from stock.

13. Identify the item below and briefly describe its purpose.

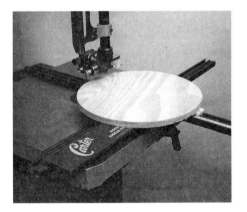

Carter Products Company, Inc.

_____
_____
_____

_____ 14. A miter jig can be used instead of a miter gauge for ____° cuts.

15. Identify the type of fences shown below.

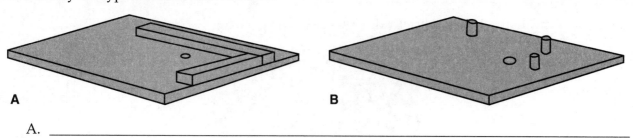

A. _____

B. _____

16. Identify the item below.

Adjustable Clamp Co.

_____

_____

_____ 17. A(n) ____ trims and fits precise 45° joints for moulding.

18. Explain why a multipurpose machine requires a great deal of planning.

_____

_____

_____

_____

_____

19. List two advantages and two disadvantages of multipurpose machinery.

_____

_____

_____

_____

Name _____

Shop Accessories, Jigs, and Special Machines

Name accessories, jigs, and special machines you would like to have in your shop and explain why.

Name _____ Date _____ Course _____

CHAPTER 39
Sharpening

Instructions: *Carefully read Chapter 39 of the text and answer the following questions.*

1. Which is more dangerous: a dull tool or a sharp tool? Explain why.

   _____

   _____

   _____

2. What are the two main considerations in achieving a sharp edge on a steel tool?

   _____

   _____

3. List three types of tools a cabinetmaker may be required to sharpen.

   _____

   _____

   _____

_____ 4. ____ help hold the tool at a fixed angle to prevent rounding the edge.

5. Explain you shouldn't let a tool get too hot in the process of grinding.

   _____

   _____

_____ 6. Which of the following statements is *not* true?
   A. Grinding is usually the first step in returning a dull edge to the desired angle and removing nicks in a steel blade.
   B. Honing further refines the scratch pattern in the tool edge.
   C. Before proper grinding can be performed a tool edge must be stropped.
   D. Final stropping of the tool edge is done for final deburring of the sharpened edge.

7. Describe the following abrasive types:

   A. Waterstones _____

   _____

   _____

   B. Diamond stones _____

   _____

   _____

   C. Oil stones _____

   _____

   _____

   D. Ceramic stones _____

   _____

   _____

   E. Abrasive sheets _____

   _____

   _____

_____ 8. The angle on the cutting face of the tool is referred to as the ____.

9. Explain what to do if your tool becomes too hot to hold while grinding.

   _____

   _____

10. List three types of tool blades that are used to hone edge tools by hand.

    _____

    _____

    _____

11. How do most motorized sharpening machines work?

    _____

    _____

12. Profile knife grinders are used to produce knives for what two tools?

    _____

    _____

Name _____

_____ 13. *True or False?* Most shops use a sharpening service for their carbide-tipped tooling.

_____ 14. *True or False?* Circular saw blades and router bits are usually sharpened in house.

15. Explain the steps you would take to sharpen a chisel that has a small nick in the cutting edge. Explain how you would know when it is sharp.

_____

_____

_____

_____

_____

Name _____

Shaping an Edge

Draw the edge shape profile of the tools listed below. Diagram angles where applicable.

1. Chisel.

2. Cabinet scraper. See Figure 24-14 in the text.

3. Hand scraper. See Figure 24-14 in the text.

4. Plane blade.

CHAPTER 40
Case Construction

Instructions: *Carefully read Chapter 40 of the text and answer the following questions.*

1. List four examples of typical cases.

   _____

   _____

   _____

   _____

_____ 2. With a(n) ____ case, you do not have to edgeband the visible edges of plywood or particleboard.

_____ 3. A(n) ____ case has no face frame; the edges of the top, bottom, and side members are exposed.

_____ 4. The 32mm System is a recent variation of ____ construction.
   A. face frame
   B. frameless
   C. frame-and-panel
   D. web frame

_____ 5. Regardless of which method of case construction you choose, any assembled case should be perfectly ____.

6. List five factors to consider when choosing materials for casework.

   _____

   _____

   _____

   _____

   _____

_____ 7. In forming large panels, which of the following is least desirable to use?
   A. Lumber
   B. Plywood
   C. MDF
   D. Particleboard

_____ 8. The case ____ refers to the top, bottom, and sides.

9. Identify the types of joints shown below.

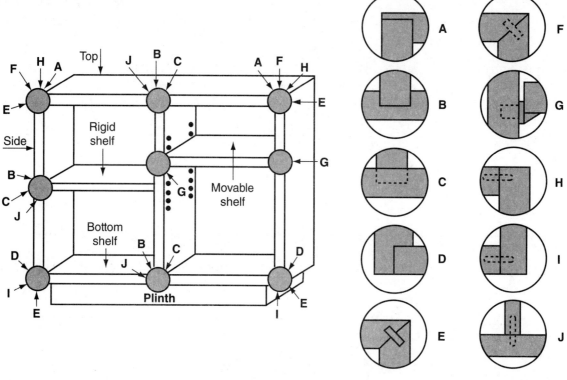

A. _____
B. _____
C. _____
D. _____
E. _____
F. _____
G. _____
H. _____
I. _____
J. _____

_____ 10. ____ divide a case into levels for storage.

_____ 11. *True or False?* Lumber shelving is less likely to warp than plywood or particleboard.

Name _____

12. Identify the types of dados used to hold fixed shelves.

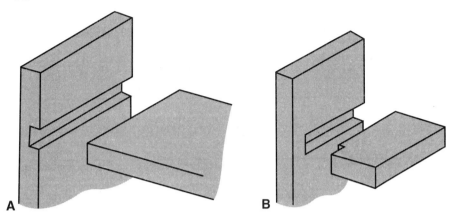

A. _____

B. _____

_____ 13. When the edge of a divider is visible on the finished case, cover it with veneer or laminate ____.

_____ 14. A(n) ____, or base, provides toe clearance on one or more sides of a case.

_____ 15. Tops on cases that open up are generally ____.

16. What is the purpose of the face frame?

_____

_____

_____ 17. Horizontal parts of the frame are called ____.

_____ 18. Vertical parts of the frame are called ____.

19. Identify types of joints used for face frames.

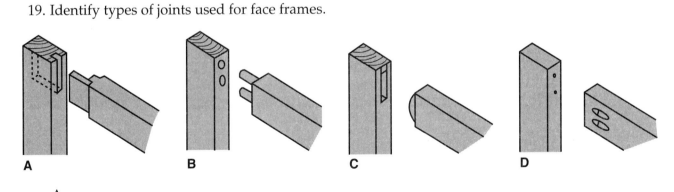

A. _____

B. _____

C. _____

D. _____

_____ 20. Successful assembly requires all except ____.
   A. joints that fit exactly
   B. components that are sized correctly
   C. a quick-setting adhesive
   D. components that are sanded smooth

21. Explain why a case should be assembled without adhesive (a dry run).

_____

_____

_____

_____ 22. ____ assembly folds to create a case.

_____ 23. The 32mm System ____.
   A. is a standard form of frameless case construction
   B. provides a modular system of case assembly and hardware installation
   C. is governed by the minimum spacing of spindles on a multiple-spindle boring machine
   D. All of the above.

_____ 24. The ____ distance provides precise alignment of hinge base plates.

_____ 25. ____ are 8 mm and accept dowels.

26. List three benefits of the 32mm System.

_____

_____

_____

_____

_____ 27. With 32mm System case construction, the vertical row of system holes are ____ mm.

_____ 28. System holes in the vertical row are spaced ____ mm apart.

_____ 29. The vertical row's system holes are spaced ____ mm from the front and rear edge of the panel.

_____ 30. Ideally, the height of case side panels should be a multiple of ____ mm, plus the material thickness.

Name _____

Casework Joints

Sketch a drawing similar to the figure in Question 9 of the workbook. Draw the joints *you* would choose for the application. Then explain why you chose each joint.

Explanation:

_____

_____

_____

_____

_____

_____

_____

_____

_____

_____

Name _____ Date _____ Course _____

CHAPTER **41**

Frame and Panel Components

Instructions: *Carefully read Chapter 41 of the text and answer the following questions.*

1. List two design purposes frame and panel components serve.

   _____

   _____

   _____

2. List five engineering purposes frame and panel components serve.

   _____

   _____

   _____

   _____

   _____

   _____

_____ 3. *True or False?* Frames must be used with panels.

_____ 4. Vertical frame side members are called ____.

_____ 5. Horizontal frame members are called ____.

_____ 6. A vertical piece other than the outside frame is called a(n) ____.

Copyright Goodheart-Willcox Co., Inc.
May not be reproduced or posted to a publicly accessible website.

7. Identify the parts indicated on the frame and panel assembly below.

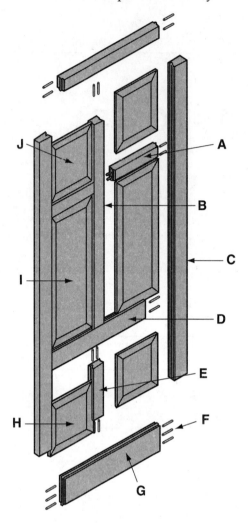

Shopsmith, Inc.

A. _____
B. _____
C. _____
D. _____
E. _____
F. _____
G. _____
H. _____
I. _____
J. _____

Name _____

8. Identify the frame and panel profiles below.

A. _____
B. _____
C. _____
D. _____
E. _____
F. _____
G. _____
H. _____
I. _____
J. _____
K. _____

9. List four questions you should ask yourself when trying to decide on what type of frame you will make.

_____
_____
_____
_____

_____ 10. A stub mortise and tenon is a popular frame joint because the panel groove also serves as the ____.

11. Name two ways stub mortise and tenon frame corners can be reinforced.

_____
_____

_____ 12. A haunched mortise and tenon does not require ____.

13. When making a profiled inside edge, what order should be used to shape the stiles, rails, and mullions?

_____

_____

_____

_____ 14. Wood or vinyl stops and plastic retainers hold panels in ____ frames.

_____ 15. Raised or beveled and raised panels provide decoration or the look of ____.

_____ 16. A(n) ____ panel is made when the surfaces of a product should be flat.

_____ 17. A(n) ____ panel lies below the surface of the frame.

_____ 18. *True or False?* It is best to apply finish to panels before installing them.

19. What type of frame construction is shown below?

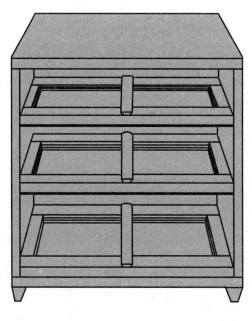

_____

Name _____

20. Identify the items indicated on the illustrations of joinery in web frame construction.

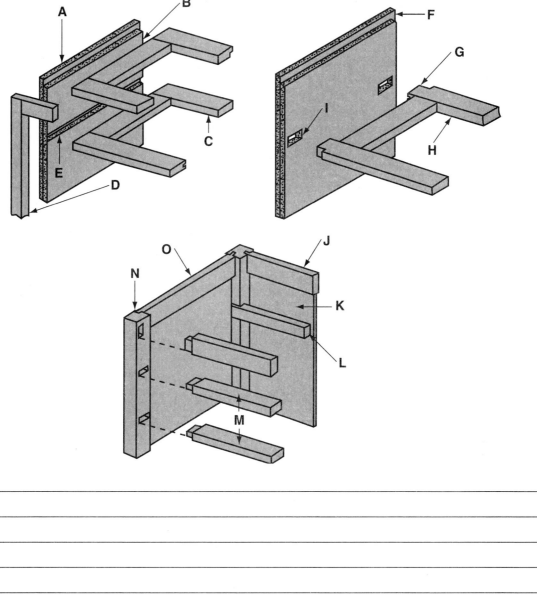

A. _____
B. _____
C. _____
D. _____
E. _____
F. _____
G. _____
H. _____
I. _____
J. _____
K. _____
L. _____
M. _____
N. _____
O. _____

Name _____

Frame and Panel Door Design

Imagine that you are designing kitchen cabinets for yourself. You have chosen to use a frame and panel type of construction for your doors. Sketch your door design and explain the tooling and procedure you would follow to make the door.

Sketch:

Explanation:

_____

_____

_____

_____

_____

_____

_____

_____

_____

_____

Name _____ Date _____ Course _____

CHAPTER 42
Cabinet Supports

Instructions: *Carefully read Chapter 42 of the text and answer the following questions.*

_____ 1. ____ raise cases or furniture above the floor.

2. What type of bracket feet are shown in the photo below?

Bracket feet

OZaiachin/Shutterstock.com

_____

_____ 3. Each side of a(n) ____ bracket foot is S-shaped.

287

4. Name the leg styles shown below.

A   B   C   D   E   F   G

Gerber

A. _____
B. _____
C. _____
D. _____
E. _____
F. _____
G. _____

5. List four basic methods of mounting legs vertically.

_____

_____

_____

_____

_____

_____ 6. Use a(n) ____, hanger bolt, and wing nut to secure removable legs to the apron.

_____ 7. ____ join a central pedestal with dowels, dovetail joints, and mechanical fasteners.

_____ 8. It is easier to join an apron to a(n) ____ section of a leg rather than a tapered section.

Name _____

9. Identify the types of decorative legs shown below.

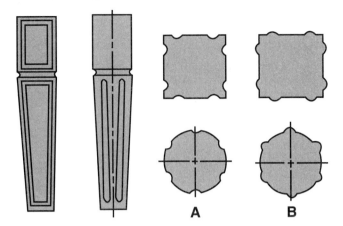

A. _____

B. _____

_____ 10. The ____ leg, a distinguishing feature of Queen Anne and French Provincial furniture, often has additions, called ears, extending from each side.

11. Identify the type of legs shown on this piece of furniture.

Laurel Crown Furniture

_____

_____ 12. Which of the following strengthens table and chair supports?
A. Stretchers
B. Rungs
C. Shelves
D. All of the above.

_____ 13. A series of shelves and spindles can be layered and assembled using ____ screws.

_____ 14. *True or False?* Posts are longer than legs.

15. Identify the cabinet support shown below.

Dyrlund

_____

16. Identify the two items below.

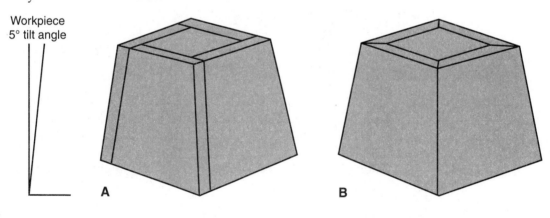

A. _____

B. _____

_____ 17. *True or False?* Having the case sides on the floor is the most complex method of support.

Name _____

18. Identify each of the items below.

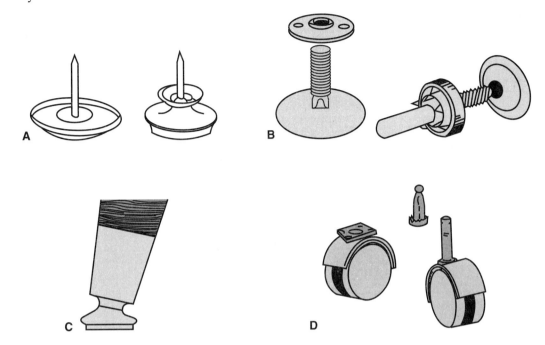

A. _____
B. _____
C. _____
D. _____

Name _____

Choose Your Support

In this chapter, you have learned about the many ways to support cabinetry and furniture. If you owned a business, you would have to choose the specific techniques and styles to use in the manufacture of your pieces. Draw thumbnail cabinet sketches of supports you would choose to use on an island cabinet, a dining chair, and a dining table. Explain how you reached your decisions.

Island cabinet _____

_____

_____

_____

_____

_____

_____

_____

Dining chair _____

_____

_____

_____

_____

_____

_____

_____

Dining table _____

_____

_____

_____

_____

_____

_____

_____

Name _____ Date _____ Course _____

CHAPTER **43**

Doors

Instructions: *Carefully read Chapter 43 of the text and answer the following questions.*

1. Name the three basic ways for doors to operate.

   _____

   _____

   _____

_____ 2. Large openings, wider than ____", usually require two hinged doors.

3. Identify the basic ways that hinged vertical doors mount to a cabinet.

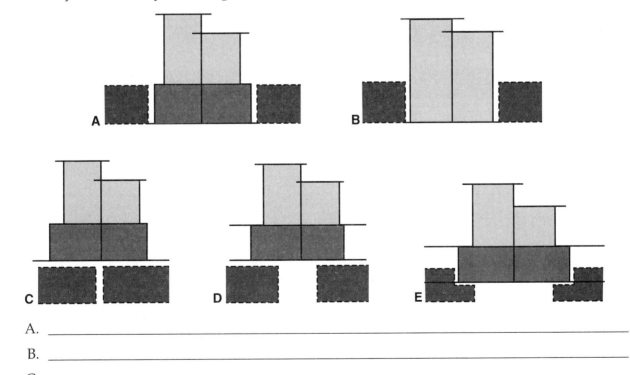

   A. _____

   B. _____

   C. _____

   D. _____

   E. _____

_____ 4. When installing hinges, make sure the pins of the hinges align perfectly to prevent the door from being ____.

_____ 5. Doors over ____' should have a third hinge to help maintain door alignment.

_____ 6. When selecting fasteners to mount doors, make sure that the screws are ____" shorter than the door or case thickness.

7. Identify the treatments on the striker edges on the double door installations below.

A   B   C   D

   A. _____
   B. _____
   C. _____
   D. _____

8. Identify the special hinge types used for mounting the flush doors below.

A (Door / Frame)   B   C

Liberty Hardware

   A. _____
   B. _____
   C. _____

Name _____

9. Describe what is being done in the photos below.

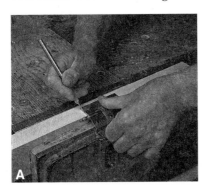

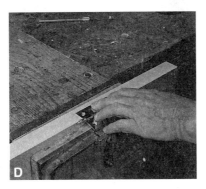

Chuck Davis Cabinets

A. _____
B. _____
C. _____
D. _____
E. _____

_____ 10. ____ doors cover some or all of the face frame, or case edges, in frameless cabinetry.

11. Identify the hinge types used to mount overlay doors.

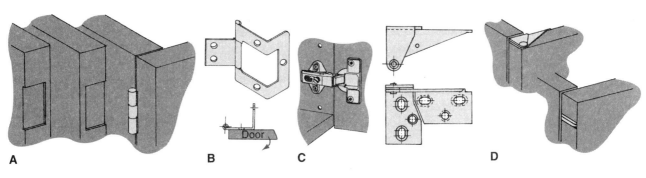

A. _____
B. _____
C. _____
D. _____

12. Identify the type of hinge shown below.

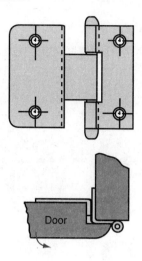

Liberty Hardware

_____

_____ 13. ____ hinges are common on lids and hinged-leaf tables.

_____ 14. Normally, unframed ____ doors rest in a U-shaped hinge leaf and are held by set screws.

_____ 15. ____ doors have two or three panels that glide past each other.

_____ 16. Sliding doors are supported and guided by a pair of metal or plastic ____.

_____ 17. The height of a sliding door must measure ____″ less than the distance between the inside of the top track and the top edge of the bottom track.

_____ 18. Horizontal sliding ____ must be in place when you fasten the second track.

_____ 19. ____ doors consist of narrow wood or plastic slats bonded to a heavy cloth backing, usually canvas.

20. What two factors determine slat size and shape in tambour doors?

_____

_____

_____ 21. For a tambour door, the tracks should be nearly twice as long as the ____.

_____ 22. Which of the following hold doors closed and also help open them?
   A. Pulls
   B. Knobs
   C. Catches
   D. Latches

Name _____

23. Describe what is happening in the photos below.

Chuck Davis Cabinets

A. _____

B. _____

C. _____

Name _____

Door Design Decisions

Contrast which type of hinged door mount configurations (See Figure 43-2) is the least expensive to make? Which would be the most expensive for the cabinetmaker to make? Explain in detail your choices and why.

Name _____ Date _____ Course _____

CHAPTER **44**

Drawers

Instructions: *Carefully read Chapter 44 of the text and answer the following questions.*

1. Name two factors that affect drawer construction.

   _____

   _____

   _____

_____ 2. *True or False?* Drawers in a chest generally are more shallow near the floor than they are at the top.

_____ 3. *True or False?* The drawer type and door mount should be the same.

4. List four ways drawer fronts fit in or against cabinets.

   _____

   _____

   _____

   _____

   _____

5. What are the five parts of a drawer?

   _____

   _____

   _____

_____ 6. Drawer fronts are usually made of ____.
   A. inexpensive, close-grain hardwood lumber
   B. the same material as the cabinet
   C. plywood
   D. hardboard

_____ 7. The length of drawer sides varies according to cabinet ____.

_____ 8. A(n) ____ drawer bottom rests in a groove cut in the back.

_____ 9. A(n) ____ bottom is cut so that the back sits on the bottom.

_____ 10. *True or False?* Sand all drawer parts after you cut the joints.

_____ 11. To assemble a drawer, apply glue to the ____ and the side-back joints.
   A. front-side joints
   B. drawer front groove
   C. drawer bottom groove
   D. All of the above.

12. What type of joints were used in the front-side and side-back joints below?

DBS drawer box specialties

_____

_____ 13. For accuracy, you should make a half-blind dovetail joint using a(n) ____.
   A. dovetail jig
   B. template
   C. router with the appropriate bit
   D. All of the above.

14. What type of dovetail joint has an overhang that butts against the face frame or case edge?

_____

_____ 15. *True or False?* Dovetails on lip drawers are shaped together.

Name _____

16. What type of joint is shown in the illustration below?

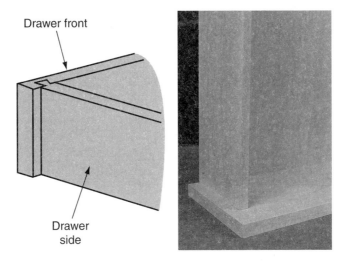

Patrick A. Molzahn

_____

17. What type of joint is shown below?

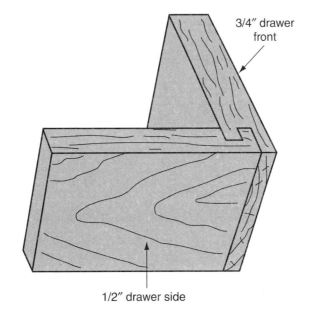

_____

18. What type of joints are shown below?

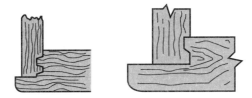

_____

19. Name the typical side-back joint shown below.

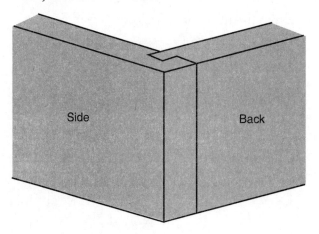

_____

20. Identify the types of drawer bottoms shown below.

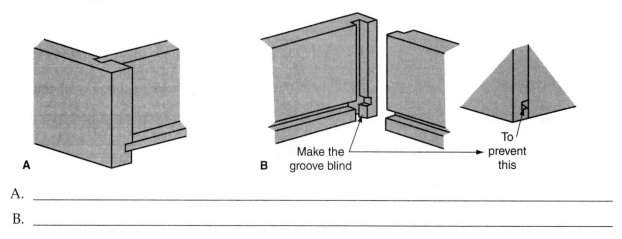

A. _____
B. _____

_____ 21. Most producers of high-quality, solid wood cabinetry use ____ construction.

22. Identify the parts indicated on the construction shown below.

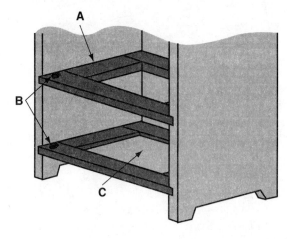

A. _____
B. _____
C. _____

Name _____

23. Look at the illustration below. What are the purposes of the kicker block and runner?

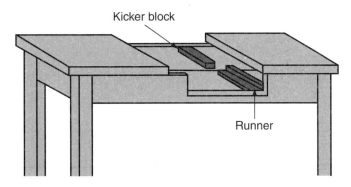

Kicker block: _____

Runner: _____

24. Identify the items indicated on the illustration below.

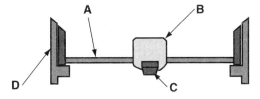

A. _____

B. _____

C. _____

D. _____

25. What are the items below and where are they installed?

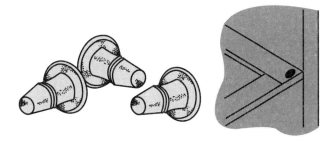

Rockler Companies, Inc.

_____

_____

26. What type of runner is shown below?

Rockler Companies, Inc.

_____

27. Why should you be cautious when selecting full extension and telescoping slides for free-standing cabinets?

_____

_____

28. List four factors that help determine the number and placement of pulls.

_____

_____

_____

_____

Name _____

Your Drawer Design

You are a cabinet shop owner. You must decide on how your drawers are going to be designed. As you know from the chapter, there are many options. Design exactly how you would build the drawers in your shop in the space below. It is your choice to draw either a view drawing (top, front, rear and one side view) or a cabinet drawing. Do not add dimensions; focus on the design configuration and materials. At the bottom of the page, add specifications of design ideas that are difficult to draw, such as specifying what type of fasteners and wood you are using.

Specifications: _____

_____

_____

_____

Name _____ Date _____ Course _____

CHAPTER 45
Cabinet Tops and Tabletops

Instructions: *Carefully read Chapter 45 of the text and answer the following questions.*

_____ 1. In addition to being attractive, the top of a cabinet or table must be ____ and functional.

_____ 2. *True or False?* Typically, the cabinet or tabletop is the first part to be assembled.

_____ 3. To prevent ____, make the top from a series of wide strips, alternating the faces.

_____ 4. *True or False?* Marble and granite are most often used for kitchen and bath countertops.

5. Identify the three edge treatments below.

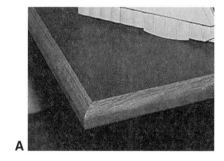

Chuck Davis Cabinets; Patrick A. Molzahn; Roth

A. _____

B. _____

C. _____

6. Identify the type of wood band shown in the illustration below.

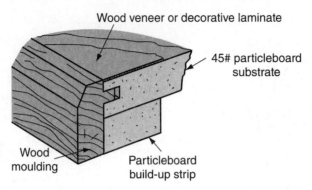

_____

7. What is edgebanding?

_____
_____

_____ 8. Which of the following plastic edgebanding types may be applied with a hand iron?
A. HPDL
B. PVC
C. Melamine
D. None of the above.

_____ 9. While metal was once very popular for kitchen countertops, the material commonly used today to create a durable edge is ____.

Name _____

10. Identify the types of hardware used for attaching tops shown below.

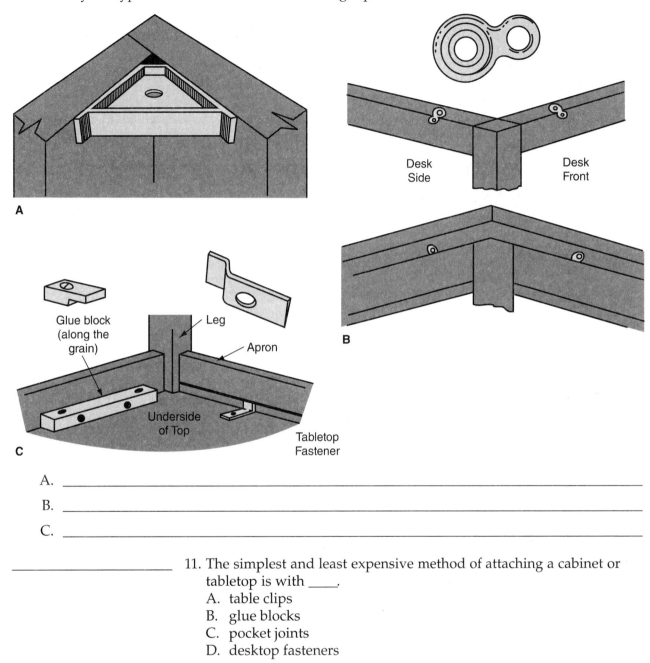

A. _____

B. _____

C. _____

_____ 11. The simplest and least expensive method of attaching a cabinet or tabletop is with ____.
   A. table clips
   B. glue blocks
   C. pocket joints
   D. desktop fasteners

_____ 12. ____ tables have one or two sections that hang when the table is not in use.

13. On the lines below, briefly describe what is happening in each step of attaching a tabletop with plate joinery.

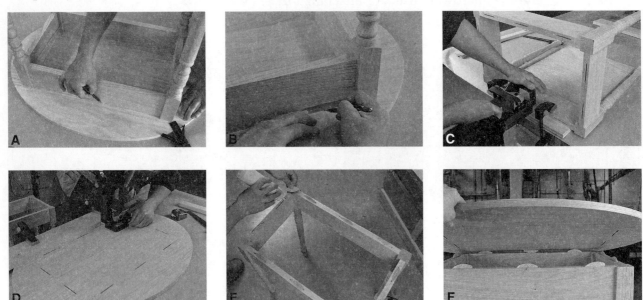

Patrick A. Molzahn

A. _____
B. _____
C. _____
D. _____
E. _____
F. _____

14. What type of joint is shown in the illustration below?

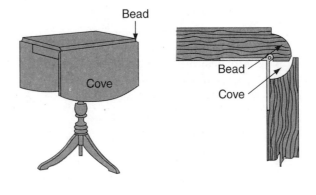

Name _____

15. Name the type of support shown in the illustration below.

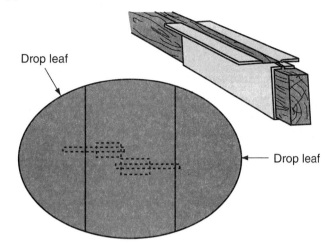

Rockler Companies, Inc.

_____

16. Name the type of support shown in the illustration below.

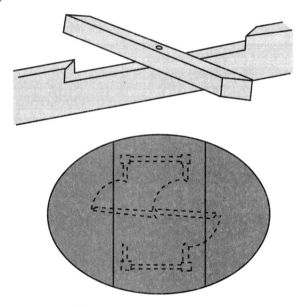

_____

17. Name the type of support shown in the illustration below.

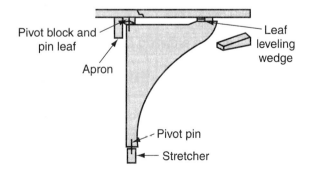

_____

18. Name the type of table illustrated below.

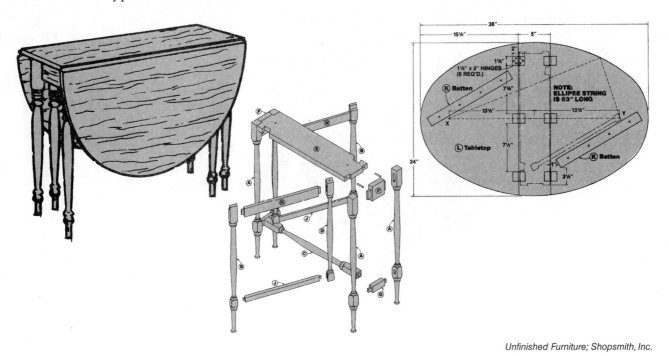

Unfinished Furniture; Shopsmith, Inc.

_____

19. Identify the item shown below.

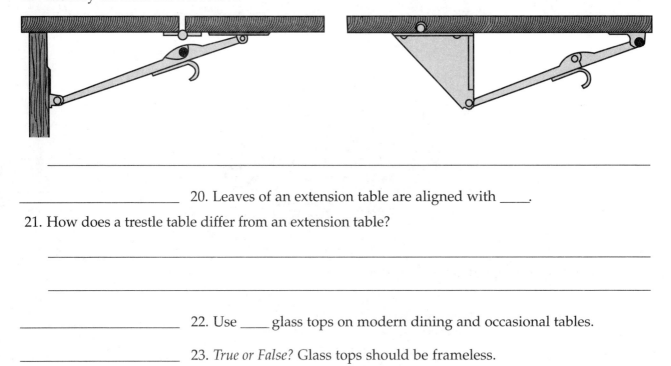

_____

_____ 20. Leaves of an extension table are aligned with ____.

21. How does a trestle table differ from an extension table?

_____

_____

_____ 22. Use ____ glass tops on modern dining and occasional tables.

_____ 23. *True or False?* Glass tops should be frameless.

Name _____

Tabletop

Your cabinet company has been given the contract to design and build tables for a restaurant. In the space provided, draw a cross section showing the materials you would use and how the tables would be designed. In the blank area below that, explain how the tops would be made. Your design should specify what materials are used as substrate, edging, backing, finish, and main covering material.

Drawing:

Procedure:

_____

_____

_____

_____

_____

_____

_____

_____

_____

_____

_____

_____

_____

Name _____ Date _____ Course _____

CHAPTER **46**
Kitchen Cabinets

Instructions: *Carefully read Chapter 46 of the text and answer the following questions.*

1. To meet kitchen needs, the cabinetmaker must consider what four factors?

   _____

   _____

   _____

   _____

2. Name eight examples of items that might be stored in kitchen cabinets.

   _____

   _____

   _____

   _____

   _____

   _____

   _____

   _____

_____ 3. ____ refer to countertop or tabletop areas where certain tasks are performed.

4. Name and briefly describe six major work centers in a kitchen.

_____
_____
_____
_____
_____
_____
_____
_____
_____
_____
_____

5. On the kitchen floor plan below, draw in the kitchen triangle.

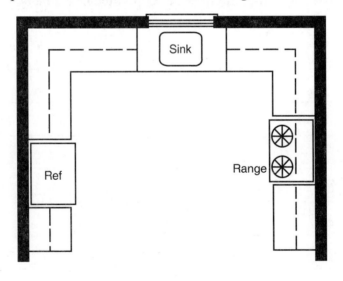

Name _____

6. What wiring and piping should be in place before cabinets are installed?

Electrical	Gas	Plumbing

_____ 7. An efficient kitchen has a triangle measuring less than _____ feet, but greater than _____ feet.

8. List five safety factors to consider when planning a kitchen.

_____
_____
_____
_____
_____
_____

9. Identify the kitchen designs shown below.

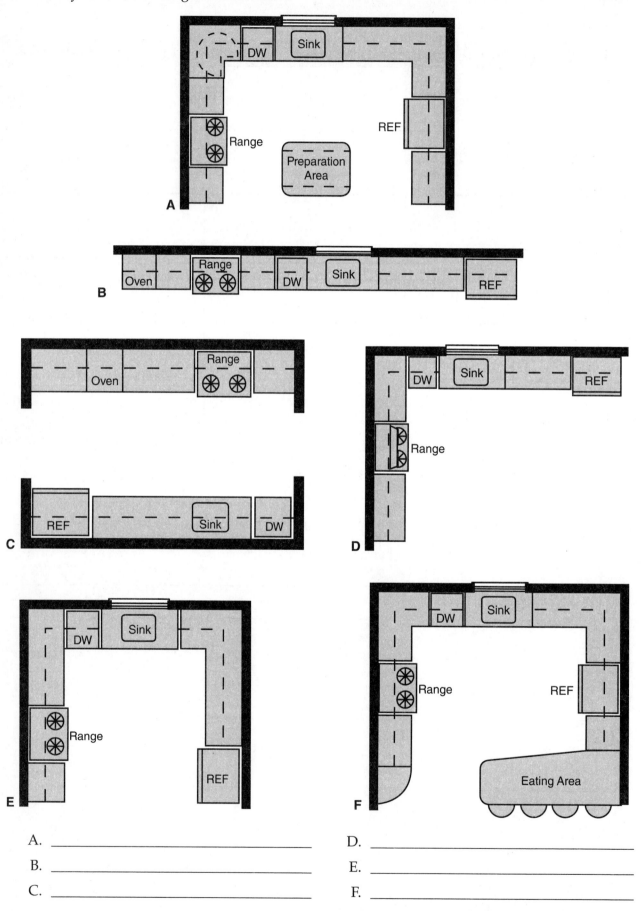

A. _____
B. _____
C. _____
D. _____
E. _____
F. _____

Name _____

_____ 10. A convenient work center has ____ space around appliances.

11. Name four cabinet accessories that can be used to help maximize storage.

_____

_____

_____

_____

12. Identify the types of kitchen cabinets indicated below.

National Kitchen and Bathroom Association

A. _____
B. _____
C. _____

_____ 13. ____ cabinets are floor units 34 1/2″ high and 24″ deep with widths from 9″ to 48″ in 3″ increments.

_____ 14. ____ cabinets are hanging units 15″–30″ high and 12″ or 30″ deep ranging in width from 12″ to 48″ in 3″ increments.

_____ 15. ____ cabinets are 83 1/2″ high units that extend from the floor to the soffit.

16. Identify the types of corner cabinets shown below.

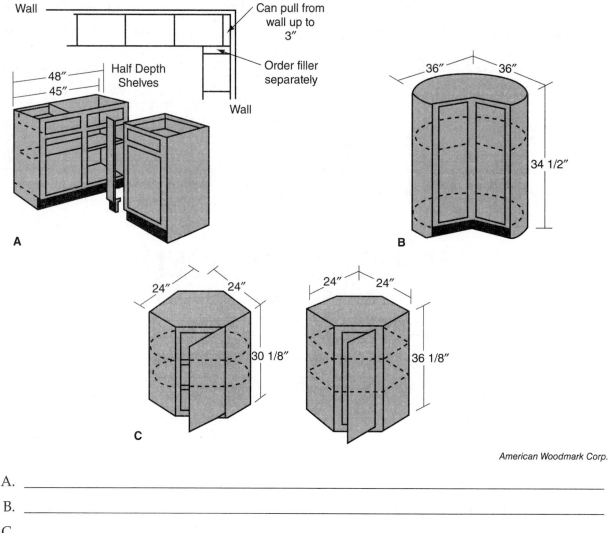

American Woodmark Corp.

A. _____

B. _____

C. _____

_____ 17. ____ cabinets can be found over islands and peninsulas.

_____ 18. ____ are open wall units for storing china, crystal, and other display items.

_____ 19. A(n) ____ fits between the countertop and wall cabinet.

Name _____

20. List six tools or types of equipment that you need to install kitchen cabinets and describe how they will be used .

_____
_____
_____
_____
_____
_____
_____
_____
_____
_____
_____
_____

_____ 21. Cabinets must be mounted ____.
    A. plumb
    B. square
    C. to wall studs
    D. All of the above.

_____ 22. Once you find one stud on each wall, locate the others by measuring ____″ left and right.

_____ 23. A(n) ____ encloses the space between the ceiling and wall cabinets.

_____ 24. *True or False?* The wall cabinet to hang first is one in a corner.

_____ 25. *True or False?* Lay out cabinet heights from the lowest point on the floor.

_____ 26. *True or False?* Begin installing base cabinets in the center.

_____ 27. Place ____ behind and under cabinets to account for irregular walls and floors.

_____ 28. The ____ is installed once the base cabinets are fully secure.

_____ 29. Most commercial and custom cabinets built today are made of ____ products such as structural particleboard and MDF.

30. The construction details for a base cabinet are shown below. Identify the name of each item specified on the bill of materials by filling in that column on the chart below.

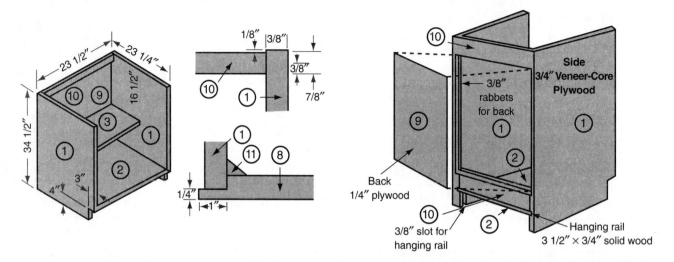

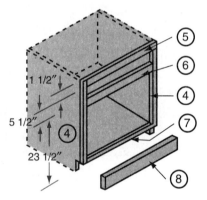

Bill of Material			
Item	Req'd.	Name	Size
1	2		3/4 × 23 1/4 × 34 1/2
2	1		3/4 × 23 1/8 × 22 3/4
3	1		3/4 × 12 × 22 3/4
4	2		3/4 × 1 1/2 × 30 1/2
5	1		3/4 × 2 × 21
6	1		3/4 × 1 1/2 × 21
7	1		3/4 × 1 × 21
8	1		3/4 × 4 × 24
9	1		1/4 × 22 3/4 × 23 1/2
10	2		3/4 × 3 1/2 × 22 3/4
11	3		3/4 × 3/4 × 3

_____ 31. ____ cabinets refer to custom units constructed on location.

_____ 32. Wall cabinets are constructed much like base units except the top is ____ into the sides.

_____ 33. All industrialized countries except the United States use ____ measurements.

Name _____

Kitchen Cabinet Checklist

Pretend you are designing a kitchen for your "dream home." Complete the checklist below to help you determine your kitchen requirements.

Checklist

Before you visit your kitchen specialist, fill out this handy checklist designed to help you determine your own personal kitchen requirements.

Cabinetry

Cabinet Style _____ Color _____
Wood Type (or Laminate) _____

Appliances

Electric or Gas Range _____ Built-in Oven _____
Electric or Gas Cooktop _____ Built-in Double Oven _____
Downdraft Cooktop _____ Microwave Oven _____
Grill _____ Built-in Oven/Microwave Combination _____
Electric or Gas Burner/Grill Combination _____ Ventilation System _____
Refrigerator _____ Trash Compactor _____
Freezer _____ Waste Disposer _____
Dishwasher _____ Other _____

Special Options

☐ Butler's Pantry   ☐ Bake Center   ☐ Special Sink   ☐ Plate Rail
☐ Snack Bar   ☐ Center Work Island   ☐ Second Sink   ☐ Other _____
☐ Appliance Garage   ☐ Range Hood   ☐ Custom Beam
☐ Planning Desk   ☐ Decorative Glass Door Inserts   ☐ Custom Appliance Panels

Accessories

(See our Accessories Guide for the complete selection of kitchen accessories.)

Wall Cabinets

☐ Microwave Cabinet   ☐ Spice Rack   ☐ Corner Tambour Cabinet   ☐ Hinged Glass Doors
☐ Open Shelf Cabinet   ☐ Corner Lazy Susan   ☐ Wall Quarter Circle Cabinet   ☐ Other _____
☐ Wall Wine Rack   ☐ Tambour Cabinet   ☐ Pigeonhole

Base Cabinets

☐ Cutlery Divider   ☐ Pull-Out Table   ☐ Base Hamper   ☐ Visible Storage Rack
☐ Tray Divider   ☐ Tote Trays   ☐ Tilt-Out Soap Tray   Baskets
☐ Bread Box   ☐ Wastebasket   ☐ Corner Lazy Susan   ☐ Other _____
☐ Pull-Out Chopping Block   ☐ Pull-Up Mixer Shelf   ☐ Adjustable Roll-Out Shelves

Tall Cabinets

☐ Tray Divider   ☐ Visible Storage Rack/Baskets   ☐ Multi-Storage Cabinet   ☐ Other _____
☐ Base Hamper   ☐ Built-in Oven Cabinet   ☐ Adjustable Roll-Out Shelves

Other Design Considerations

Sink and Fixtures _____ Type of Wall Coverings _____
Special Window Treatment _____ Space Arrangement Changes _____
Type of Flooring _____
Type of Countertop _____ Special Design Needs _____
Type of Lighting _____

Quaker Maid

Measuring Appliances

Visit an appliance store or the appliance department of a department store. Complete the chart below. (You may wish to bring along a tape measure.)

Appliance	Brand/Make	Model Number	Width Left to Right	Depth Front to Back	Height
Freestanding Range					
Drop-In Range					
Countertop Cooktop					
Built-In Wall Oven					
Microwave					
Built-In Dishwasher					
Built-In Trash Compactor					
Refrigerator					
Sink					
Other					
Other					

American Woodmark Corp.

Name _____ Date _____ Course _____

CHAPTER **47**
Built-In Cabinetry and Paneling

Instructions: *Carefully read Chapter 47 of the text and answer the following questions.*

_____ 1. Storage units, often custom-made to fit special space restrictions, are called ____.

2. Give six examples of where added storage might be needed in a home.

_____
_____
_____
_____
_____
_____
_____

3. Give three examples of where added workspace might be needed in a home.

_____
_____
_____

4. List five general rules that you should follow when designing built-in storage.

_____
_____
_____
_____
_____
_____

_____ 5. *True or False?* Custom-made bathroom vanities can follow the guidelines and standard sizes given for kitchen cabinets, but are higher.

_____ 6. Which of the following can be utilized for storage space?
   A. Attics
   B. Stairwells
   C. Wall units
   D. All of the above.

7. Which three built-in activity areas need utilities?

_____

_____

_____

8. List seven utility and storage factors that must be considered when designing a room to support laundry activities.

_____

_____

_____

_____

_____

_____

_____

_____ 9. Plan for ____ to prevent moisture buildup.

_____ 10. ____ is an alternative to plastered or gypsum board-covered walls.

11. Identify the three types of millwork paneling shown below.

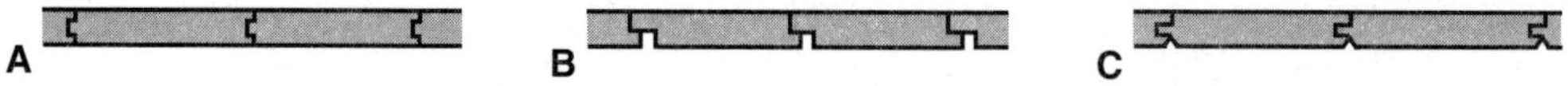

California Redwood Association

A. _____

B. _____

C. _____

Name _____

_____ 12. *True or False?* When installing paneling, mouldings should be added over joints.

_____ 13. Horizontal millwork paneling attaches directly to ____.

_____ 14. Place horizontal ____ strips if you will be installing vertical millwork paneling.

15. List the steps necessary to estimate the number of sheets of paneling that will be needed for a project.

_____
_____
_____
_____
_____
_____
_____
_____
_____

_____ 16. When installing vertical paneling, leave about ____" at the floor.

_____ 17. *True or False?* Begin installing horizontal paneling at the ceiling with the tongue edge up.

_____ 18. Begin installing ____ panels in a corner.

_____ 19. *True or False?* Millwork paneling is generally much easier to install than sheet paneling.

_____ 20. *True or False?* Install backing ____ vertically for vertical sheet paneling.

21. After you press paneling with adhesive to the wall, what are two reasons to pull it away for a short time?

_____
_____

22. Identify the steps used to install paneling.

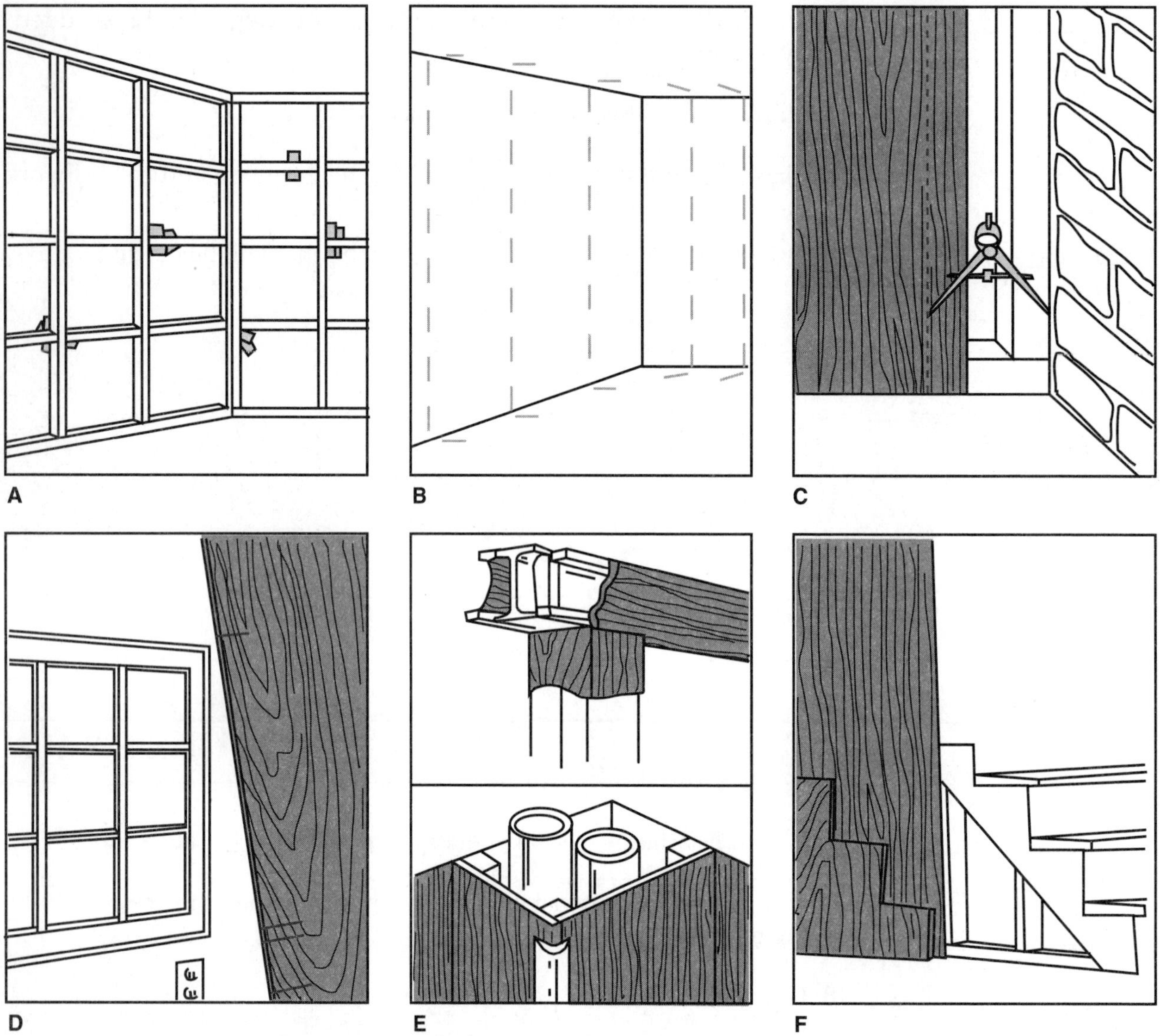

A. _____
B. _____
C. _____
D. _____
E. _____
F. _____

Name _____

23. Identify the panel installations shown below.

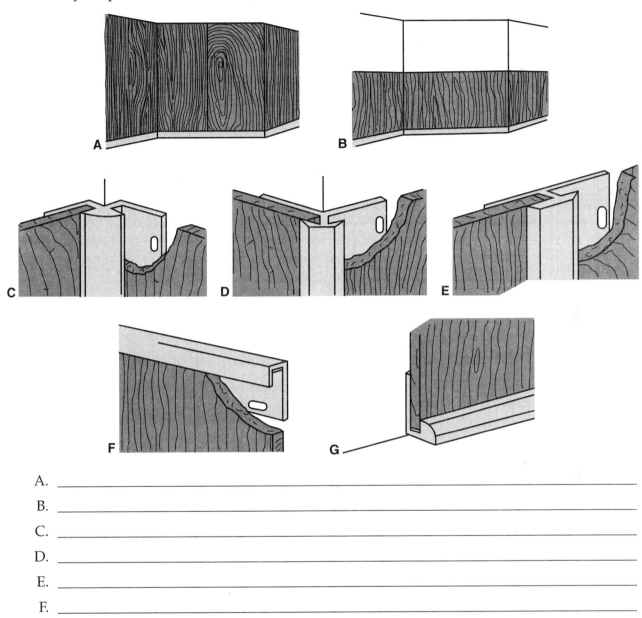

A. _____
B. _____
C. _____
D. _____
E. _____
F. _____
G. _____

_____ 24. Trim and moulding should be installed with ____ nails.

Name _____

Add Built-in Cabinets to Your Home

Find a good place in your home to add built-in cabinets or shelves. Draw a plan view of the room (looking down from the ceiling) and an elevation view (eye level looking straight at the wall where you would add the built-in cabinets or shelves).

Plan View:

Elevation View with Addition:

Name _____ Date _____ Course _____

CHAPTER **48**

Furniture

Instructions: *Carefully read Chapter 48 of the text and answer the following questions.*

_____ 1. Always process duplicate parts of desks ____.
  A. as needed
  B. at the same time
  C. one at a time
  D. None of the above.

_____ 2. When making a clock, choose the desired movement ____.
  A. after the clock is designed
  B. just before finishing the outside
  C. during the design stage
  D. None of the above.

_____ 3. Few chairs are stable if the legs are ____.
  A. perpendicular to the seat
  B. at an angle to the seat
  C. at different angles between the front and back
  D. None of the above.

4. List three parts of a typical bed.

_____

_____

_____

_____ 5. ____ support the box springs and mattress.

6. A chart of standard mattress sizes is given below. In the Type column, write in the correct type of mattress.

Bedding Sizes—in Inches (millimeters)			
Type	Cotton or foam mattresses and box springs	Water bed	
		Hard-side	Soft-side
		48 × 84 (1220 × 2134)	
	36 × 74 (914 × 1880)		36 × 74 (914 × 1880)
	54 × 74 (1372 × 1880)		54 × 74 (1372 × 1880)
	60 × 80 (1524 × 2032)	60 × 84 (1524 × 2134)	60 × 80 (1524 × 2032)
	76 × 80 (1930 × 2032)	72 × 84 (1829 × 2134)	76 × 80 (1930 × 2032)
	18 × 36 (457 × 914)		
	22 × 39 (559 × 991)		
	23 × 46 (584 × 1168)		
	25 × 51 (635 × 1295)		
	27 × 51 (686 × 1295)		
	31 × 57 (787 × 148)		
	33 × 66 (838 × 1676)		
	36 × 76 (914 × 1930)		

Name _____

7. Identify the parts indicated on the illustration below.

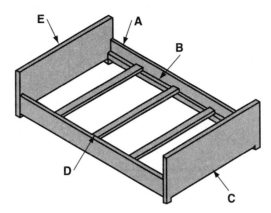

A. _____

B. _____

C. _____

D. _____

E. _____

_____ 8. A(n) ____-side water bed mattress is simply a vinyl inner tube that must be placed within a solid frame.

9. Name two ways water beds can be supported.

_____

_____

_____ 10. The ____ protects the wall, provides decoration, and sometimes offers storage.

_____ 11. ____ prevent bedspreads from slipping off the bed and often provide decoration.

_____ 13. A(n) ____ mirror is supported in a stand.

_____ 14. ____ are used to separate work areas, reduce drafts, or provide privacy.

15. Look at the illustration below. It is an example of _____ furniture.

Unfinished Furniture

_____

Name _____

Selecting Furniture

In the space provided, mount photos, pictures from the newspaper, magazines (make sure to get approval before cutting up a magazine or journal) furniture ads of furniture you'd like to have in your dream bedroom.

Name _____ Date _____ Course _____

CHAPTER 49
Finishing Decisions

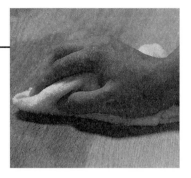

Instructions: *Carefully read Chapter 49 of the text and answer the following questions.*

1. List four safety precautions to take when applying finishing materials.

_____

_____

_____

_____

_____

_____

_____

_____

_____ 2. Finishing involves applying the proper ____ materials to assembled products.

_____ 3. A finish protects wood, wood products, and metal from ____.
    A. moisture
    B. harmful substances
    C. wear
    D. All of the above.

Copyright Goodheart-Willcox Co., Inc.
May not be reproduced or posted to a publicly accessible website.

4. Complete the following chart by filling in the sections under the Finish heading.

Finishing Selected Wood Species

This chart recommends finishing practices for popular softwoods and hardwoods. The table suggests procedures for finishing these woods with opaque or clear coatings. Stains may be optional (O). Filler may be required (R) to produce a flat surface on some open grain species.

Species	Wood Type			Finish				Filler
	Softwood	Hardwood		Penetrating	Opaque	Build-up	Stain	
		Open	*Closed*					
Alder, red			●					
Ash		●						R
Banak	●							
Basswood			●					
Beech			●					
Birch			●					
Butternut		●						R
Cedar, Aromatic Red	●							
Cherry			●					
Chestnut		●						R
Cottonwood			●					
Cypress, bald	●							
Ebony			●					
Elm, American		●						R
Fir, Douglas	●							
Gum, red			●					
Hackberry		●						R
Hickory		●						
Lauan (Phillipine Mahogany)		●						R
Limba		●						R
Mahoganies, genuine		●						R
Mahogany, African		●						R
Maple, hard			●					
Maple, soft			●					
Oak, red or white		●						R
Paldao		●						R
Pecan			●					
Pine, Ponderosa	●							
Pine, sugar	●							
Pine, yellow	●							
Primavera		●						R
Redwood	●							
Rosewood			●					
Santos Rosewood (Pau Ferro)			●					
Sapele		●						R
Sassafras		●						R
Satinwood			●					
Spruce	●							
Sycamore			●					
Teak			●					
Tulip, American (Yellow Popular)			●					
Walnut, American		●						R
Willow			●					
Zebrawood			●					

Name _____

5. List the five steps involved in finishing.

_____

_____

_____

_____

_____

_____ 6. Unwanted natural coloring (mineral deposits, blue stain, etc.) can be removed by ____.

_____ 7. Which of the following involves spattering the surface of wood with black enamel before applying the final topcoat?
A. Scorching
B. Distressing
C. Imitation distressing
D. None of the above.

8. List five methods for applying coating materials.

_____

_____

_____

_____

_____

9. List the four steps in applying topcoating.

_____

_____

_____

_____

_____ 10. ____ changes the color of the wood.

_____ 11. ____ contain dyes and resins that are almost totally absorbed into the wood.

_____ 12. ____ contain insoluble powdered colors that bond to the wood surface with resins.

_____ 13. A(n) ____ is a thinned coat of sealer that often is applied before stain.

_____ 14. _____ is a liquid or paste material that is used to fill pores.

_____ 15. A(n) _____ is a thin, clear coat that fills the pores of the wood and serves as a base coat.

_____ 16. _____ prepares surfaces for opaque coatings.

_____ 17. Which of the following is *not* a penetrating finish?
    A. Linseed oil
    B. Phenolic resin oil-based coatings
    C. Shellac
    D. Wax

18. Name one advantage and one disadvantage of using penetrating finishes.

_____

_____

_____

_____

_____ 19. _____ finishes do not penetrate the wood; they form a film on the surface.

20. Describe the difference between gloss built-up finish and satin built-up finish.

_____

_____

_____

_____

_____

_____

Name _____

21. Complete the clear penetrating coatings chart below by filling in the names of the topcoatings described.

Topcoating	Solvents or Thinners	Applications	Advantages	Disadvantages
Natural				
	Mineral spirits or Turpentine	Table legs Lamps Picture frames Boxes	Resists moderate heat Soft Dull finish	Slow drying Yellows with age Requires frequent rubbing with oil
	Mineral spirtis or Turpentine	Small cabinets	Resists moderate heat Hard Glossy or satin Stable color Durable Easy to apply	May need to be recoated if surface is marred
	None needed	Any	Extra protection from moisture	Provides minimal protection from stains and chemicals
Synthetic				
	Mineral spirits	Paneling	Glossy or satin Inexpensive Easy to apply	Damaged by alcohol, solvents, and food acids
	Turpentine	Tabletops	Glossy More expensive than alkyds Easy to apply	Highly resistant to common household chemicals

22. Complete the built-up topcoatings chart below by filling in the names of the topcoatings described.

Topcoating	Solvents or Thinners	Applications	Advantages	Disadvantages
Natural				
	Alcohol	Cabinets Furniture Tables	Durable Scratches easy to hide	Surface should be kept dry Not resistant to water
	Lacquer thinner	Cabinets Furniture Tables	Hard Various gloss levels Scratches easy to hide Resists chemicals	Toxic during application
	Turpentine or Mineral spirits	Cabinets Furniture Tables	Durable Various gloss levels May contain stain for coloring and protecting	Yellows over time
Synthetic				
	Read container label	Same as natural finishes	Chemical curing for toughness Some clean up with soap and water Less yellowing than natural finishes Various gloss levels	
	Lacquer thinner	Same as natural finishes	Improved over natural Various gloss levels	Same as natural

_____ 23. ____ topcoatings are the least reflective.

24. Name two stages of the drying time of a topcoating.

_____

_____

_____ 25. ____ are ink transfers bonded to the surface.

_____ 26. ____ involves buffing the surface with wax and, generally, is the last step in the finishing process.

_____ 27. Which of the following provides a visual effect of texturing while the surface remains smooth?
   A. Mottling
   B. Antiquing
   C. Gliding
   D. Marbelizing

_____ 28. Usually, clear lacquer is applied to hardware and other metal products to prevent ____.

_____ 29. Which of the following finishes can be removed with alcohol?
   A. Shellac
   B. Lacquer
   C. Enamel
   D. Varnish

30. In the space below, list the steps you would take to produce a stained, clear satin finish on a cabinet made of an open grain hardwood.

_____

_____

_____

_____

_____

_____

_____

_____

_____

_____

Name _____

Decisions! Decisions! Decisions!

Making the right decisions when it comes to what finishing method to use on a particular wood piece will come with experience. The information found in this chapter can speed up that process. Using that information, decide how you would finish the pieces listed here. Use the charts provided in the text and refer back to chapters that describe the particular type of wood used to make the piece. Defend your decisions and contrast them with some of the methods that you did not use. In other words, not only tell why you chose a particular method but tell why you did not choose another.

Small, aromatic cedar chest _____

_____

_____

_____

_____

_____

_____

Cabinet made of maple veneer plywood and solid wood face frame _____

_____

_____

_____

_____

_____

_____

Solid walnut coffee table _____

_____

_____

_____

_____

_____

_____

Name _____ Date _____ Course _____

CHAPTER **50**

Preparing Surfaces for Finish

Instructions: *Carefully read Chapter 50 of the text and answer the following questions.*

_____ 1. The method used to prepare a surface depends on the final ____.

2. List at least four safety precautions to take when preparing to apply a finish.

_____

_____

_____

_____

_____

_____ 3. *True or False?* Remove surface defects after applying a finish.

_____ 4. Dents can often be raised with ____.
   A. abrasives
   B. a drop of water
   C. stain
   D. None of the above.

5. Name four products that can repair chips, scratches, and voids.

_____

_____

_____

_____

_____ 6. ____ consists of real wood flour in a resin and works well for small defects.

_____ 7. ____ are cellulose fiber fillers.

_____ 8. *True or False?* Glue and wood dust mixed together make an effective patch for large cracks and holes.

_____ 9. Precolored ____, made of wax, fills small holes or cracks.

_____ 10. ____ is melted to fill defects in wood.

11. Identify the tool below.

Patrick A. Molzahn

_____

12. Why do some cabinetmakers recommend filling defects for opaque finishes after the first coat?

_____

_____

_____ 13. Excess dry glue should be removed with ____.
   A. a hand scraper or cabinet scraper, followed by adhesive paper
   B. water
   C. your fingernail
   D. a chisel

_____ 14. Oil spots can be dissolved and removed with ____ or lacquer thinner.

_____ 15. ____ is helpful in removing mildew, blue stain, and clamping stains.

_____ 16. ____ is the process of swelling wood fibers and any last dents, dimples, or other pressure marks.

_____ 17. *True or False?* Higher density MDF is the easiest to work with when finishing.

Name _____

18. What is occurring in the photo below?

Patrick A. Molzahn

19. What is occurring in the photo below?

Patrick A. Molzahn

20. What is occurring in the photo below?

Patrick A. Molzahn

Name _____

What Tools to Use?

You are in the business of making fine custom cabinets for homes. Make a list of the supplies and tools you must have on hand to prepare the surfaces for the finish you apply. Explain how you would use each item.

Name _____ Date _____ Course _____

CHAPTER 51
Finishing Tools and Equipment

Instructions: *Carefully read Chapter 51 of the text and answer the following questions.*

1. Name five finish application practices.

   _____
   _____
   _____
   _____
   _____

2. List three guidelines to consider when selecting a finish application method.

   _____
   _____
   _____

_____ 3. Applying finishes with brushes ____.
   A. involves a limited amount of equipment compared to spraying
   B. is fast compared to spraying
   C. Both A and B.
   D. None of the above.

4. List four features to consider when selecting brushes.

   _____
   _____
   _____
   _____

_____ 5. *True or False?* Brushes with synthetic bristles are suitable for applying shellac and lacquer.

_____ 6. Brushes with natural bristles are suitable for applying ____.
   A. oil-based finishes
   B. water-based finishes
   C. Both A and B.
   D. None of the above.

7. Identify the parts of the paintbrush below.

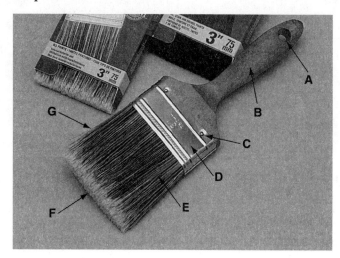

Stanley Tools

A. _____
B. _____
C. _____
D. _____
E. _____
F. _____
G. _____

_____ 8. To dip a brush in finish is called ____.

_____ 9. When dipping a brush into finish, cover the bristles ____.
   A. more than 60% of the length
   B. no more than 50% of the length
   C. the entire length
   D. None of the above.

_____ 10. ____ refers to applying an even film of finish.

11. Describe the proper ways to clean and store a paintbrush.

_____
_____
_____
_____
_____

Chapter 51 Finishing Tools and Equipment **359**

Name _____

_____ 12. With ____, the finish is atomized into tiny droplets that flow from the gun nozzle, producing a flat, nearly flawless film.

13. Identify the equipment shown below.

Hitachi Power Tools U.S.A., Ltd.

_____

_____ 14. ____ guns do not have a valve to shut off airflow.

_____ 15. ____ guns allow the user to control the air and liquid flow rates.

16. Identify the types of nozzles shown below.

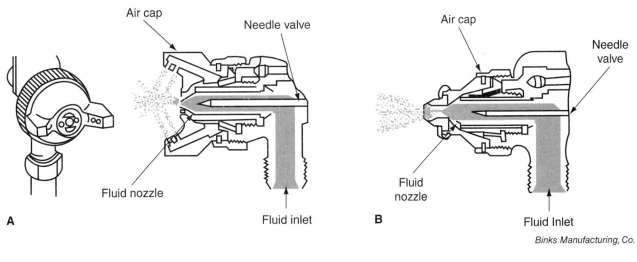

Binks Manufacturing, Co.

A. _____

B. _____

17. Various spray pattern troubles when using an air spray system are listed below along with their probable causes. Determine the remedies that should be used.

Spray Pattern Trouble	Probable Cause	Remedy
Bottom or top heavy pattern.	Debris in gun nozzle.	
Pattern heavy on side.	One hole in external-mix air cap is clogged.	
Pattern heavy in center.	Material too thick. Too little air pressure. Fluid tip too large for film being sprayed.	
Split pattern.	Partially clogged fluid tip.	

18. Various gun performance troubles when using an air spray system are listed below along with their probable causes. Determine the remedies that should be used.

Gun Performance Trouble	Probable Cause	Remedy
Constant fluttering spray.	Loose fluid tip. Low on fluid in cup. Gun and cup tipped too much. Clogged fluid tube.	
(suction gun)	Fluid too thick. Clogged cup lid vent. Leaky fluid tube or needle packing. Leaky nut or seal around fluid cup top.	
Fluid leaks at packing nut.	Loose packing nut and hard or defective packing.	
Fluid leaks at nozzle.	Needle not seating.	
No fluid flow.	Out of material. Clogged spray system.	
No fluid flow from pressure gun.	Lack of air pressure in cup or pot. Leaky gasket on cup/gun connection or pot. Clogged internal air hole in cup or pot.	
No fluid flow from suction gun.	Dirty air cap and fluid tip. Clogged air vent. Leaky gun connections.	

Name _____

19. Various work appearance troubles when using an air spray system are listed below along with their probable causes. Determine the remedies that should be used.

Work Appearance Trouble	Probable Cause	Remedy
Sags and runs	Dirty gun nozzle. High fluid pressure. Fluid too thick or thin. Gun too close or at wrong angle to work. Slow gun movement. Improper triggering by starting or stopping over work.	
Streaks	Dirty gun nozzle. Air pressure too high. Gun too far from or at wrong angle to work. Gun moved too rapidly.	
Excessive fog (overspray)	Air pressure too high. Fluid pressure too low. Material too thin. Gun too far from work. Spray beyond work at end of pass.	
Rough surface (orange peel)	Low air pressure. Material too thick. Poor quality thinner or poorly mixed and strained. Gun too close, too far, or moved too rapidly. Overspray strikes tacky surface.	

20. Explain how to use the wet film thickness gauge, and explain why it is important.

_____

_____

_____

_____

21. On the spray gun below, indicate where lubrication would be needed.

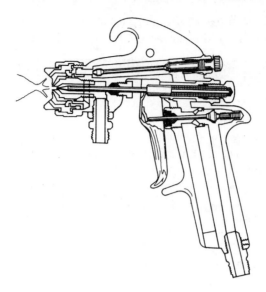

ITW DeVilbiss

_____ 22. With _____ spraying, a fluid pump forces finish up into the gun nozzle where high pressure causes atomization.

23. List three advantages of airless equipment over air-spray equipment.

_____

_____

_____

Name _____

24. Identify the parts of the commercial airless spray system.

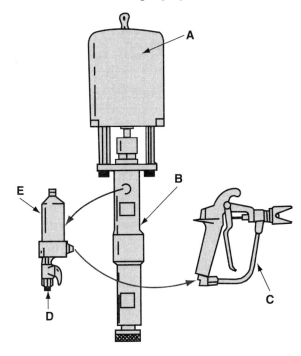

Binks Manufacturing, Co.

A. _____
B. _____
C. _____
D. _____
E. _____

25. Various spray pattern troubles when using an airless spray system are listed below along with their possible causes. Determine the remedies that should be used.

Spray Pattern Trouble	Cause	Remedy
Excessive fogging (overspray).	Sprayer too far from surface. Material too thin. Pressure too high. Worn spray tip.	
Material sputtering or uneven flow.	Cup empty. Sprayer at angle. Material too thick. Intake tube drawing in air. Control knob set improperly. Debris in nozzle. Oversize or worn tip. Worn valve.	

(Continued)

Spray Pattern Trouble	Cause	Remedy
No material flow.	Pump motor not running. Pump not operating. Loose or damaged intake tube. Liquid too thick. Filter screen clogged. Pressure control knob closed. Valve not opening. Clogged tip.	
Tails in spray pattern.	Inadequate fluid delivery. Fluid not atomizing.	
Hourglass pattern.	Inadequate fluid delivery. Plugged or worn tip.	
Surging.	Pulsating fluid delivery.	
Runs and sags.	Sprayer movement too slow. Spray too close to product. Excess material flow. Liquid too thin.	

_____ 26. ____ are a handy alternative to using bulky and inconvenient spray equipment for finishing smaller products.

_____ 27. ____ is a simple coating practice used to apply stain and penetrating finishes.

_____ 28. ____ is done to apply an even layer to small components quickly.

29. List four examples of industrial finishing equipment.

_____

_____

_____

_____

30. List at least four safety precautions to take when applying coating materials.

_____

_____

_____

_____

Name _____

Finish a Piece in Your Garage

You have just completed a small trinket box working with hand tools in your garage. In keeping with this same idea of simplicity, you want to finish this box using the simplest and safest methods. Explain what materials and tools you would need and how you would proceed with the process of applying a finish to your trinket box.

_____
_____
_____
_____
_____
_____
_____
_____
_____
_____
_____
_____
_____
_____
_____
_____
_____
_____
_____
_____
_____
_____
_____
_____

Set Up a Finish Department

You are the owner of a growing successful mill and cabinet company. You have been sending all of your cabinets out to be stained and finished but now you would like to set up your own finishing department. In the space below, draw a simple floor plan of your finish room and label the finishing equipment you would use and the locations of the specific finishing operations. Pay attention to product flow through the department. In the space following your drawing, describe your operation.

_____

_____

_____

_____

_____

_____

_____

_____

_____

_____

Name _____ Date _____ Course _____

CHAPTER 52
Stains, Fillers, Sealers, and Decorative Finishes

Instructions: *Carefully read Chapter 52 of the text and answer the following questions.*

1. What does preparing a surface for topcoating involve?

   _____
   _____
   _____
   _____
   _____
   _____

_____ 2. A(n) ____ is a coat of thinned sealer that helps control stain penetration.

_____ 3. ____ alters a wood's color, accents grain patterns, or hides unattractive grain.

4. Characteristics of various stains are given in the chart below. Fill in the blanks with the name of the appropriate type of stain.

	Natural				Synthetic		
	_____	_____	_____	_____	_____	_____	_____
Purpose	All hardwoods and softwoods.	All hardwoods and softwoods.	All lumber and veneer.	Lumber, plywood, wood products.	Darken sap streaks and hiding scratches.	Lumber and manufactured wood panels, not plywood.	Lumber and manufactured wood panels, not plywood.
Advantages	Excellent color that is permanent. Low cost, nonfading. Clean with soap and water.	Quick drying. Penetrates deeply.	Penetrates evenly and deeply. Excellent color. Convenient to apply.	Penetrates evenly. Good color that is nonfading.	Very fast drying. Shallow penetration. Excellent for sapwood and shading.	Colors well, especially on rough wall panels.	Colors well, especially on rough wall panels.

(Continued)

	Natural				Synthetic		
Disadvantages	Raises grain. Could soften water-base adhesives. Penetrates quickly, thus can show overlap marks.	Dries too fast to brush.	Fades in sunlight. Tends to bleed unless properly sealed.	Fades in sunlight. Shallow penetration. Slower drying. Difficult to touch up. Overlap shows.	Tends to fade and bleed. Difficult to apply evenly.	Bleeds with some natural wood resins. Visible nails will rust.	Bleeds with some natural wood resins. Visible nails will rust.
Application	Spray Brush Wipe	Spray	Brush Spray Wipe	Brush Spray Wipe	Brush on sapwood. Spray for shading.	Brush Roll Spray Wipe	Brush Roll Spray Wipe
Solvent	Water	Alcohol Acetone	Mineral spirits Turpentine Naptha	Mineral spirits Turpentine	Denatured alcohol Acetone	Water	Mineral spirits
Relative cost	Low	High	Medium	Medium	High	Low	Medium
Grain raising	Bad	Very little to none	None	None	Very little	Bad	None
Grain clarity	Excellent	Excellent	Excellent	Good	Good	Some	Some
Bleeding	None	Very little to none	Bad	None	Bad	Little	Little
Fading	None	None	Some	None	Some	Some	Some
Drying time to recoat	1 to 4 hours	10 minutes to 3 hours	1 to 4 hours	3 to 12 hours	10 to 15 minutes	2 to 4 hours	2 to 4 hours
Color source	Water soluble aniline dyes	Alcohol soluble aniline dyes	Oil soluble aniline dyes	Pigments and sometimes dyes	Alcohol soluble aniline dyes	Pigment	Pigment

_____ 5. It is wise to stain a(n) ____ or hidden area on a cabinet first.

_____ 6. *True or False?* Water stain tends to raise the wood grain.

7. What is one disadvantage of non-grainraising (NGR) stain?

_____

_____

_____ 8. *True or False?* Oil stains penetrate more deeply than water stains.

_____ 9. ____ stains are used in furniture restoration and for evening the tone between sapwood and heartwood.

_____ 10. *True or False?* Latex stain is recommended for high-quality cabinetmaking.

_____ 11. ____ stain is similar to latex or pigment stains in that it hides the surface.

Name _____

12. List six tips to follow when staining.

_____

_____

_____

_____

_____

_____

_____ 13. ____ is the process of packing a paste material into large pores of open grain woods, leveling the surface in preparation for topcoating.

_____ 14. The most common filler is ____.

_____ 15. *True or False?* Filler can be applied before, after, or along with stain.

_____ 16. *True or False?* Large pores in wood require a thinner filler mix than do small pores.

17. List the 11 steps involved in applying filler.

_____

_____

_____

_____

_____

_____

_____

_____

_____

_____

_____

18. Give two reasons sealer is needed on wood surfaces.
_____
_____

19. List four types of sealers.
_____
_____
_____
_____

_____ 20. Which method provides the best results when applying sealers to wood?
   A. Brushing
   B. Rubbing
   C. Spraying
   D. Dipping

_____ 21. The sealing coat on metal is usually thinned clear ____ used to keep the topcoat adhere better.

_____ 22. Priming materials must be compatible with the ____.

_____ 23. When priming metal products, apply a colored primer that will ____ with the topcoat color.

24. List eight decorative finishes that can be applied to cabinetry and furniture.
_____
_____
_____
_____
_____
_____
_____
_____

_____ 25. Which of the following involves outlining and highlighting edges and other areas with gold paint?
   A. Antiquing
   B. Gilding
   C. Shading
   D. Graining

Name _____

26. List three safety rules for applying stains, fillers, sealers, and decorative finishes.

_____

_____

_____

Name _____

Choose a Decorative Finish

After reading the chapter, choose the decorative finish you found the most interesting. What kind of project would you apply it to? Explain what materials you would need and the process you would follow to apply the finish to your project.

_____
_____
_____
_____
_____
_____
_____
_____
_____
_____
_____
_____
_____
_____
_____
_____
_____
_____
_____
_____
_____
_____
_____

Name _____ Date _____ Course _____

CHAPTER **53**

Topcoatings

Instructions: *Carefully read Chapter 53 of the text and answer the following questions.*

_____ 1. ____ is the final protective film on a complete product.

_____ 2. ____ coatings hide the surface and are applied to manufactured wood products and poor quality lumber.

3. Name three environmental factors that must be considered when applying a topcoating.

_____

_____

_____

4. What is the primary advantage of penetrating finishes over built-up films?

_____

_____

_____ 5. *True or False?* Raw linseed oil requires more drying time than boiled linseed oil.

6. List the six steps involved in applying a linseed oil finish.

_____

_____

_____

_____

_____

_____

_____

7. List the seven steps involved in applying a tung oil finish.

_____
_____
_____
_____
_____
_____
_____

_____ 8. ____ is a combination of oil and wax, and possibly stain.

_____ 9. *True or False?* A phenolic resin finish is more durable than an alkyd resin coating.

_____ 10. Which of the following is a low-build coating?
  A. Varnish
  B. Shellac
  C. Enamel
  D. Polyurethane

11. Why should you check the expiration date on shellac container labels?

_____
_____

_____ 12. The ____ process consists of applying a combination of shellac blended with other ingredients and oils.

_____ 13. The ____ in synthetic lacquer gives the dried film added flexibility.
  A. resin
  B. plasticizer
  C. solvent
  D. None of the above.

_____ 14. *True or False?* Varnishes sold under the names "spar" or "marine" are extra-tough products.

_____ 15. In ____ varnish, the resin is rosin.

_____ 16. Which of the following is the most wear-resistant of synthetic varnishes?
  A. Acrylic varnish
  B. Epoxy varnish
  C. Polyurethane varnish
  D. Phenolic varnish

Name _____

_____ 17. *True or False?* It is best to shake varnish to mix the ingredients.

_____ 18. At the proper viscosity, varnish spreads easily by ____.

_____ 19. *True or False?* Most enamels today are natural.

_____ 20. ____ is by far the most used oil-based wood coating.

21. List eight characteristics of synthetic water-based enamels.

_____
_____
_____
_____
_____
_____
_____
_____

_____ 22. Apply enamel as you would apply ____.
　　A. varnish
　　B. shellac
　　C. lacquer
　　D. None of the above.

_____ 23. Most built-up finishes require ____ or more coats of finish.

_____ 24. Between coats, the surface should be rubbed with a(n) ____ grit or finer abrasive.

_____ 25. The gloss of shellac, lacquer, and varnish can be changed by ____.

_____ 26. What are two ways of making a nonscratch surface?

_____ 27. Liquid removers are most appropriate for ____ surfaces.

28. Complete the chart below by filling in the remedy column.

Surface Defect	Probable Cause	Remedy
Bleeding (stain color works to surface)	Improper sealer applied.	
Blistering (bubbling of dry film)	Wet film exposed to excessive heat. Excess film thickness.	
Blushing (clear film cloudiness)	High humidity during application.	

(Continued)

Surface Defect	Probable Cause	Remedy
Fish eyes (small round imperfections)	Silicone or oil on surface.	
Crazing (film cracks in dry film)	Topcoat material too thick. Impurities on the surface.	
Flat Spots (areas of low sheen)	Uneven sealer coat.	
Orange Peel (textured surface)	Topcoat material too thick. Not enough material applied to properly flow out.	
Pinholing (small holes in dry film)	Too thick of a coat.	
Runs and Sags (irregular wet or dry film thickness)	Excessive material applied to surface. Too much solvent in material.	
Specks or Nibs (small, hard rough spots in wet or dry film)	Dirty equipment—brush, roller, spray gun, wiping cloth. Dust in air when applying film.	
Spotty Setting (uneven evaporation of solvent in wet film)	Poor surface preparation. Contamination of surface.	
Sweating (change in gloss level of dry film)	Topcoating applied over uncured sealer.	
Tackiness (slow setting)	Filler, stain, or sealer not dry before coating. Poor ventilation. Temperature too low.	
Withering (loss of dry film gloss)	Surface not sealed or not sealed evenly. Sealer not dry before topcoating.	
Wrinkling (uneven film surface)	Uncured seal coat. Incompatible products used.	

29. Why should paint remover be brushed in only one direction?

30. List five safety precautions to take when working with finishes and solvents.

Name _____

Choose a Topcoating

Listed below are several pieces. Describe what topcoating you would use. Explain in detail why you chose that topcoating, and contrast your decision with the topcoatings you did not choose.

1. An Adirondack chair made of cedar that is placed outside all year round in direct sunlight.

   _____
   _____
   _____
   _____
   _____
   _____
   _____

2. An island made of ash installed in a kitchen.

   _____
   _____
   _____
   _____
   _____
   _____
   _____

3. A paint-grade cabinet made of MDF.

   _____
   _____
   _____
   _____
   _____
   _____
   _____

Name _____ Date _____ Course _____

SECTION PROJECT **1-1**

Class PowerPoint Presentation

Before starting your study of cabinetmaking, it is important to understand the scope of the cabinetmaking industry. This project will help you build a foundation of knowledge about the industry.

Resource Chapters

Chapter 1—Introduction to Cabinetmaking

Chapter 6—Components of Design

Objectives

After completing this project, you should be able to:

- Understand the elements that make up the cabinetmaking industry.
- Understand how to use Microsoft PowerPoint to make a presentation.

Materials and Resources

- Computer with Microsoft PowerPoint
- Scanner

Activity Procedure

1. Read an assigned section of the chapter as an individual or in a team.
2. Scan pictures from textbook and other sources.
3. Create slides.
4. Put your name on each slide you create. The slides of your entire class will be combined into one PowerPoint presentation.
5. Present PowerPoint as a class, and present your specific section.

Final Project Requirements

- Your PowerPoint slides
- Class presentation
- Follow-up questions

Copyright Goodheart-Willcox Co., Inc.
May not be reproduced or posted to a publicly accessible website.

Follow-Up Questions

1. What were your main challenges in completing this activity?

   _____

   _____

   _____

2. What main idea stands out to you in this activity?

   _____

   _____

   _____

3. How do you feel you did on your presentation?

   _____

   _____

   _____

Name _____

Rubric for Class PowerPoint Presentation
Student Guidelines

Oral presentation   Possible points _____   Score _____

Did you speak with confidence?

_____

Slide readability   Possible points _____   Score _____

Is your main idea obvious? Is the mix of text and images appropriate?

_____

Information accuracy   Possible points _____   Score _____

Is your information accurate?

_____

Subject coverage   Possible points _____   Score _____

Is your subject coverage thorough?

_____

Follow-up questions   Possible points _____   Score _____

Were all questions answered thoughtfully?

_____

　　　　　　　　　　　　　　　Total possible points _____   Score _____

Name _____ Date _____ Course _____

SECTION PROJECT **1-2**

Mini Flip Chart of Professional Organizations

Professional cabinetmaking organizations help their members stay up-to-date on the latest developments in the industry and techniques being used. Prepare a directory mini flip chart of the professional organizations that relate to cabinetmaking professionals.

Resource Chapters

Chapter 3—Career Opportunities

Chapter 4—Cabinetmaking Industry Overview

Objectives

After completing this project, you should be able to:

- Use the mini flip chart to organize information.
- Discuss at least six cabinetmaking organizations.

Materials and Resources

- Computer with a word processing program and Internet access for conducting research on cabinetmaking organizations
- Four sheets of 8 1/2" × 11" paper

Activity Procedure

1. Conduct your research. Search using phrases such as "cabinetmaking professional organizations," "cabinetmaking supply organizations," and "architectural millwork organizations."

2. Collect four sheets of paper. Measure and make a small mark on one end of each piece, 1/2" from the bottom edge of the paper.

3. Slide the sheets so that just the 1/2" is exposed.

4. Fold the papers so that you get eight 1/2" flaps.

5. Staple twice at the fold point. See diagram on next page.

6. On the flaps (the 1/2" exposed area), write the organization names you are researching.

7. Describe the organization under the flap. Include contact information, websites, e-mail addresses, and a brief synopsis of the services the organization performs.

Final Project Requirements

- Completed mini flip chart

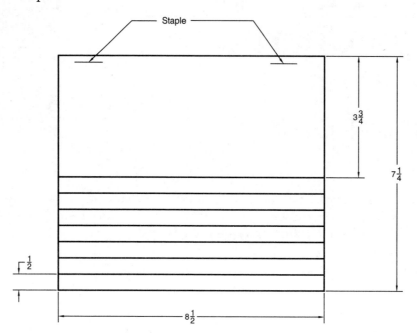

Follow-Up Questions

1. What were your main challenges in completing this project?

_____

_____

_____

_____

2. Explain how this project will be used.

_____

_____

_____

_____

3. Which organization would you join first? Explain why.

_____

_____

_____

_____

Name _____

Rubric for Mini Flip Chart of Professional Organizations
Student Guidelines

Format and appearance Possible points_____ Score_____

Is your flip chart attractive? Is it neat and straight?

_____

Number of organizations Possible points_____ Score_____

How many organizations are included in your flip chart?

_____

Information accuracy Possible points_____ Score_____

Are all organizations related to woodworking, cabinetmaking, and millwork?

_____

Follow-up questions Possible points_____ Score_____

Did you answer the questions thoughtfully?

_____

Total possible points _____ Score_____

Name _____ Date _____ Course _____

SECTION PROJECT 1-3
Identifying Shop Hazards

Everyone should contribute to maintaining a safe atmosphere in the shop. This activity gives you a chance to be a part of the safety program in your shop. Identify any hazards you see in the shop and prepare a sign warning others of the hazards.

Resource Chapters
Chapter 2—Health and Safety

Chapter 6—Components of Design

Objectives
After completing this project, you should be able to:
- Communicate about hazards in the shop.
- Provide a detailed explanation of your chosen safety rule.
- Demonstrate use of the basic elements of design.
- Demonstrate resourcefulness and creativity.

Materials and Resources
- Graph paper
- Poster board
- Markers and colored pencils

Activity Procedure
1. Brainstorm ideas.
2. Decide on final design and sketch it on a sheet of 8 1/2″ × 11″ graph paper.
3. Lay out your design in pencil on poster material.
4. When you are satisfied with the proportions of the lettering and the elements, darken and color.
5. Present your poster to your class.
6. Hang your poster in the shop.

Copyright Goodheart-Willcox Co., Inc.
May not be reproduced or posted to a publicly accessible website.

Final Project Requirements

- Preliminary sketches
- Final poster design
- Follow-up questions

Follow-Up Questions

1. What were your main challenges in completing this project?

   _____
   _____
   _____
   _____

2. Explain how this project will be used.

   _____
   _____
   _____
   _____

3. What safety concept did you decide to use for your poster?

   _____
   _____
   _____
   _____

4. Where did you hang your poster in the shop? Explain why.

   _____
   _____
   _____
   _____

Name _____

Rubric for Identifying Shop Hazards
Student Guidelines

Sketches Possible points_____ Score_____

Did brainstorming lead to good choices for your design? Is your sketch a detailed, accurate description of the final poster design?

_____

Resourcefulness Possible points_____ Score_____

Did you use time and materials wisely?

_____

Creativity and layout Possible points_____ Score_____

Did you use materials creatively and follow the elements of design to communicate your idea?

_____

Visual communication Possible points_____ Score_____

Is your main idea visible from across the room? Is 80% of remaining information visible from eight feet?

_____

Information accuracy Possible points_____ Score_____

Is the information on the poster correct and complete?

_____

Craftsmanship Possible points_____ Score_____

Did you follow the plans? Is your work neat and creative?

_____

Follow-up questions Possible points_____ Score_____

Did you answer the questions thoughtfully?

_____

Total possible points _____ Score_____

Name _____ Date _____ Course _____

SECTION PROJECT **2-1**
Project Management

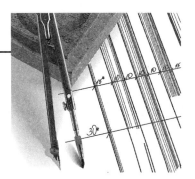

Almost all companies hire project managers. A project manager manages all the resources needed to complete a project. Project managers are the key to having materials and workers on a job at the right time and for the right price. An objective of this program is for you to develop these project management skills. The documents in this activity are tools to help you manage all the projects in this workbook.

For this activity, you are going to apply the documents to a project either you or your instructor will choose. The best approach is to choose a project that does not have a drawing and complete the drawings and the planning for that project.

Resource Chapters

Chapter 9—Production Decisions

Chapter 10—Sketches, Mock-Ups, and Working Drawings

Chapter 11—Creating Working Drawings

Chapter 12—Measuring, Marking, and Laying Out Materials

Objectives

After completing this project, you should be able to:

- Describe the elements of project management.
- Make sketches of a project to scale using the graph paper provided in this project.
- Use a plan of procedure to manage the steps and your time in this project.
- Use the bill of materials sheet to figure the total cost of a project.
- Use the 4′ × 8′ layout plan and cut list form to plan the cutting process on 4′ × 8′ panel materials.
- Use the cut work sheet to create a material cut list.

Materials and Resources

- Project management sheets included in this activity

Activity Procedure

1. Study the completed project management sheets for the trinket box included in this workbook in Section 5.

2. Draw up the project using the graph paper sheets. Include appropriate pictorial sketches, multiview drawings, detail drawings, and section view drawings, complete with dimensions and notes. You should have a minimum of three pages in your drawing set. Make the drawings to scale by establishing a certain measurement to equal a square, for example, 1" equals one square. An object 3" inches tall would be three squares tall on the graph paper.

3. If sheet materials are to be used, use the 4' × 8' layout plan and cut work sheet to plan cuts.

4. Fill out the bill of materials sheet. This should include all materials, as if you were going to the supply store to get these materials. In most cases, you cannot buy exact sizes of sheet material. Therefore, your bill of materials will not have the exact sizes of panel goods on it, but will instead simply reference how many sheets are required.

5. Complete the plan of procedure. The form included in this project has space to keep track of time spent on each procedure. Shade in when the each step starts and when it ends. This helps you, the project manager in this instance, plan when you are going to need the shop's resources.

Final Project Requirements

- Shop drawing of the project you designed
- Layout plans for the 4' × 8' sheets
- Bill of materials for the project
- Cut work sheet for solid materials (ask your instructor if you should include the 4' × 8' panel materials you placed on that sheet)
- Plan of procedure for the project

Name _____

DRAWING SCALE _____ = _____ SQUARES   DATE ___/___/___   FILE No. _____

NAME: _____

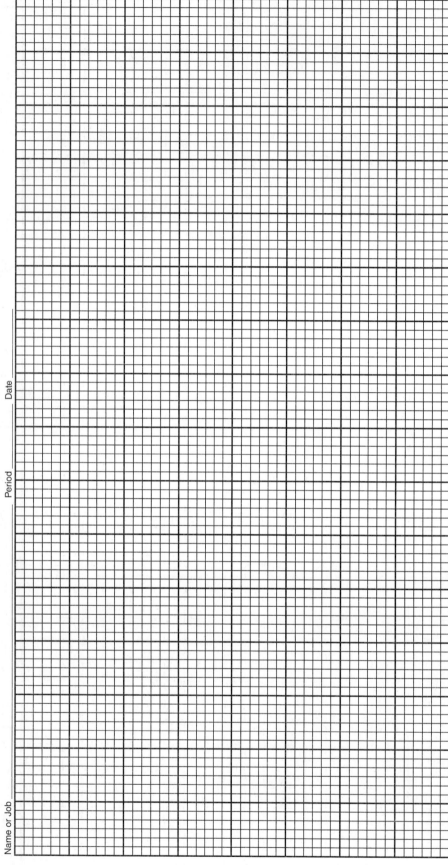
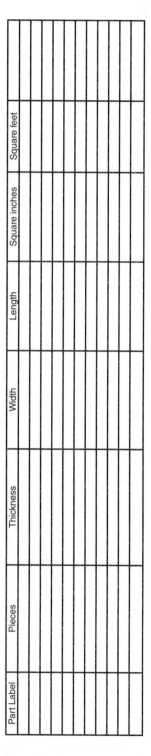

Name _____

Material Cut Sheet

Part Name	Material	Number Pieces	Rough Measurements			Cut Process Completed	Finish Measurements			Cut Process Completed
			Thickness	Width	Length		Thickness	Width	Length	

Copyright Goodheart-Willcox Co., Inc.
May not be reproduced or posted to a publicly accessible website.

Bill of Materials

Project Name: _____

Quantity	Units	Material Description	Price/Unit	Extended Price
			Subtotal	
			Tax	
			Total	

Copyright Goodheart-Willcox Co., Inc.
May not be reproduced or posted to a publicly accessible website.

Name _____

Plan of Procedure

Minutes, Hours, Days, Weeks, or Months

1 2 3 4 5 6 7 8 9 10 11 12 13 14 15 16 17 18 19 20 21 22 23 24 25

Steps

1 2 3 4 5 6 7 8 9 10 11 12 13 14 15 16 17 18 19 20 21 22 23 24 25 26 27 28 29 30 31 32

Follow-Up Questions

1. What were your main challenges in completing the process plan?

2. Explain how this plan of procedure is used.

3. Explain in your own words what is involved in project management.

4. If you leave out a step on the process plan and realize it later, what would you do to correct the plan?

5. Explain how you use graph paper to draw objects to scale.

Name _____

Rubric for Project Management
Student Guidelines

Shop drawings   Possible points_____   Score_____

Are dimensions applied correctly and drawn to scale? Are all necessary detail present?

_____

Bill of materials   Possible points_____   Score_____

Are materials figured correctly? Are all materials listed and calculated?

_____

Plan of procedure   Possible points_____   Score_____

Are essential steps listed in order? Are there time estimates? Are actual time records kept?

_____

4′ × 8′ layout plan and cut list   Possible points_____   Score_____

Was sheet used to determine the number of sheets needed on the materials bill?

_____

Cut work sheet   Possible points_____   Score_____

Was cut work sheet filled out in detail?

_____

                                     Total possible points _____   Score_____

Name _____ Date _____ Course _____

SECTION PROJECT **2-2**
Student Desk and Desk Chair Design

In this project, you will design a simple student desk and desk chair. Choose a cabinetry style that you like. You will draw shop drawings, including pictorial and multiview drawings to scale on graph paper, showing overall dimensions. The human factors should guide all of your dimensions.

Resource Chapters

Chapter 5—Cabinetry Styles

Chapter 6—Components of Design

Chapter 7—Design Decisions

Chapter 8—Human Factors

Chapter 10—Sketches, Mock-Ups, and Working Drawings

Chapter 11—Creating Working Drawings

Objectives

After completing this project, you should be able to:
- Demonstrate an understanding of cabinetry styles by explaining what influenced your choice of style in the chosen design.
- Demonstrate an understanding the components of design by explaining how each of the components were considered in the design.
- Demonstrate an understanding of the design decision process by documenting the process.
- Demonstrate an understanding of the human factors in the final design.
- Demonstrate a mastery of preliminary sketching for a design.
- Demonstrate an understanding of the different methods of pictorial sketching by creating an isometric, cabinet oblique, or perspective sketch of the chosen design.
- Demonstrate an understanding of a two-view multiview drawing by making a two-view shop drawing of the final design on graph paper.

Materials and Resources

- Textbook
- Graph paper
- Sharp pencils

Activity Procedure

1. Make rough sketches of your desk and chair. Decide exactly what you want. For instance, do you want drawers? Shelves?
2. Consider human factors in designs. Collect all the measurements you need, such as chair height and desk leg clearance.
3. Consider what style is going to influence your design.
4. Make several different thumbnail sketches of possible designs that meet the specifications you have chosen.
5. Choose and refine your design by making more detailed pictorial sketches.
6. Draw to scale detailed two- or three-view multiview drawings on graph paper.
7. Make necessary detail drawings of parts and assemblies that cannot be easily seen in the multiview drawing.

Final Project Requirements

- A list of the human factors you considered in your design.
- A statement of your design style choice.
- A statement of how the components of design were used in the design of your products.
- A statement explaining how the design process was used to arrive at the final design.
- Preliminary sketches drawn on plain paper.
- A refined pictorial sketch done on graph paper to scale of each item.
- A two- or three-view drawing of each item.
- At least one sketch on graph paper of a detail drawing and section drawing showing complicated construction detail.

The Decision-Making Process

1. Identify needs and wants. Specify what problems this product must solve.

_____

_____

_____

_____

_____

Name _____

2. Gather information. What is the current style, if any, that should be matched? Who will be using the product? List the human factors.

_____

_____

_____

_____

_____

_____

3. Create at least six preliminary ideas. Sketch each idea.

4. Pick three of your ideas as possible final designs, and refine them.

   Idea 1

   Idea 2

   Idea 3

5. Analyze the function, strength, cost, and appearance of your ideas.

   Positive aspects of Idea 1 _____

   _____

   _____

   Negative aspects of Idea 1 _____

   _____

   _____

   _____

Name _____

Positive aspects of Idea 2 _____

_____

_____

_____

Negative aspects of Idea 2 _____

_____

_____

_____

Positive aspects of Idea 3 _____

_____

_____

_____

Negative aspects of Idea 3 _____

_____

_____

_____

6. Make a decision. Choose your final idea based on the analysis you have just made. Consider the final style, drawer arrangement, doors, and proportions of each item. Explain your decision.

_____

_____

_____

_____

_____

_____

7. On graph paper, draw detailed shop drawings needed to build the final design. Your drawing set should include a separate drawing of the desk and chair, and detail or section drawings of each drawer and door with at least one section and detail included in the drawing set.

Rubric for Student Desk and Desk Chair Design
Student Guidelines

Completeness   Possible points_____ Score_____

Are there the necessary number of drawings and views to fully explain the solution and decision?

_____

Style   Possible points_____ Score_____

Do the drawings represent a specific style?

_____

Dimensions   Possible points_____ Score_____

Are extension lines, dimension lines and either tic marks, arrowheads, or dots to end dimension lines used properly?

_____

View accuracy   Possible points_____ Score_____

Do views on the same page align? Do they accurately represent the final design?

_____

Decision-making form   Possible points_____ Score_____

Does the work represent thoughtful planning? Does it follow the process?

_____

     Total possible points _____ Score_____

Name _____ Date _____ Course _____

SECTION PROJECT **2-3**

Components of Design Poster

An understanding of the components of design is an important part of making desirable designs. Design a poster that communicates the four design principles of harmony, repetition, balance, and proportion. The poster design should follow the components of design.

Resource Chapter

Chapter 6—Components of Design

Objectives

After completing this project, you should be able to:

- Demonstrate an understanding of scale.
- Demonstrate use of the basic components of design as referenced in Chapter 6.
- Demonstrate resourcefulness and creativity.

Materials and Resources

- 8 1/2" × 11" graph paper
- 18" × 24" poster board
- Miscellaneous poster and graphic materials, such as construction paper and markers.

Activity Procedure

1. Sketch ideas on a piece of 8 1/2" × 11" graph paper.
2. Be detailed in this planning stage. Include decisions on color and materials. For instance, if a purple construction paper cutout is to be included on the poster, label that on the sketch. Someone else should be able to look at your sketch and create the poster.
3. Decide on size and boldness of your print. Remember the idea here is to communicate clearly. The final poster will be twice the size of the sketch, so all elements on the sketch will be drawn half size.
4. Create the final poster on 18" × 24" poster board.

Final Project Requirements

- Preliminary scale sketches on 8 1/2" × 11" graph paper
- Final poster

Copyright Goodheart-Willcox Co., Inc.
May not be reproduced or posted to a publicly accessible website.

Follow-Up Questions

1. What were your main challenges in completing this project?
   _____
   _____
   _____

2. Explain how this project will be used.
   _____
   _____
   _____

3. What element of the design concept do you feel you understand the best? Please explain.
   _____
   _____
   _____

Name _____

Rubric for Components of Design Poster
Student Guidelines

Sketch   Possible points_____   Score_____

Was the sketch a detailed, accurate description and graphic representation of the final poster?

_____

Resourcefulness   Possible points_____   Score_____

Did the student use all materials available in a useful manner?

_____

Creativity and layout   Possible points_____   Score_____

Did creative use of materials and the elements of design accomplish the target goal of communicating an idea?

_____

Accuracy of information   Possible points_____   Score_____

Is information on the poster correct and complete?

_____

Craftsmanship and neatness   Possible points_____   Score_____

Was design well executed?

_____

Visual communication   Possible points_____   Score_____

Is the main idea visible from across the room? Is 80% of remaining information visible from eight feet?

_____

                                    Total possible points _____   Score_____

Copyright Goodheart-Willcox Co., Inc.
May not be reproduced or posted to a publicly accessible website.

Name _____ Date _____ Course _____

SECTION PROJECT **3-1**
Wood Samples Stringer

Having a sample of all the materials that are available in your shop can be very helpful when designing pieces or deciding what material to choose for your next project.

Resource Chapters

Chapter 15—Cabinet and Furniture Woods

Chapter 16—Manufactured Panel Products

Chapter 23—Sawing with Stationary Power Machines

Objectives

After completing this project, you should be able to:

- Explain the properties of all the materials used in the cabinet shop.
- Explain what units materials are sold in (sheets, board foot, square foot, etc.).
- List the materials used in your cabinetmaking shop.

Materials and Resources

- A 2″ × 5″ sample of each type of material used in your shop
- Computer with word processing software (optional)
- Laser printer (optional)

SAFETY NOTE
Before proceeding with this activity, you should have passed a general shop safety rule test, all specific safety tests on the tools used in this activity, and be certified by the instructor to use the tools and equipment needed for this activity.

Activity Procedure

1. Work with your classmates to cut solid wood material, plywood, manufactured panel materials, plastic laminate, and plastic acrylic sheets into 2″ × 5″ sample pieces.

2. Make labels that include the following information:

   Weight per unit

   Price per unit

Copyright Goodheart-Willcox Co., Inc.
May not be reproduced or posted to a publicly accessible website.

Description of properties of the material

Standard uses of the material

3. Drill a 1/4" hole in one end of the each piece, 1/2" from the end.
4. Tie pieces together using a piece of 1/8" braided nylon rope. String the rope through the holes and tie a knot, forming a loop. Use enough rope so that you can easily flip through your samples.

Final Project Requirements

- Sample stringer with the number of labeled samples assigned by your instructor

Follow-Up Questions

1. What were your main challenges in completing this project?

   _____
   _____
   _____
   _____

2. Explain how this project will be used.

   _____
   _____
   _____
   _____

3. Did you finish your project on time?

   _____

4. What piece was your favorite piece or interested you the most?

   _____
   _____
   _____
   _____

5. How many samples did you create?

   _____

Name _____

Rubric for Wood Samples Stringer
Student Guidelines

Description label Possible points_____ Score_____

Were descriptions of units, weight per unit, and material uses adequate?

_____

Number of samples Possible points_____ Score_____

Did you complete the assigned number of samples?

_____

Craftsmanship Possible points_____ Score_____

Are mill marks sanded out, corners broken properly, and holes drilled consistently in each piece?

_____

Efficiency Possible points_____ Score_____

Was the assignment finished on time?

_____

Follow-up questions Possible points_____ Score_____

Were all questions answered in a thoughtful manner?

_____

Total possible points _____ Score_____

Copyright Goodheart-Willcox Co., Inc.
May not be reproduced or posted to a publicly accessible website.

Name _____ Date _____ Course _____

SECTION PROJECT **3-2**
Moisture Content Material Testing

Moisture content of wood is a serious concern to a cabinetmaker. In this project, you will determine the moisture content of five wood samples.

Resource Chapter

Chapter 13—Wood Characteristics

Objectives

After completing this project, you should be able to:

- Identify and describe open grain wood and closed grain wood.
- Differentiate between radially (quarter sawn) and tangentially (plain sawn) samples.
- Explain how to determine moisture content.
- Use technical terms from the chapter to describe the samples.

Materials and Resources

- Digital scale
- Magnifying glass
- Five different wood species samples, one cut radially, another tangentially
- Oven

SAFETY NOTE
Before proceeding with this activity, you should have passed a general shop safety rule test and all specific safety tests on the tools used in this activity, and be certified by the instructor to use the tools and equipment needed for this activity.

Activity Procedure

1. Cut one 3/4" × 2" × 4" piece from each of the different wood species that are quarter sawn.
2. Cut one 3/4" × 2" × 4" piece from each of the different wood species that are plain sawn.
3. Review the categories on the Lab Sheet and enter your findings.
4. Using an oven and following the instruction in the chapter, dry the wood samples to zero moisture content.

5. Use a magnifying glass to inspect the grain. Determine if it is open or closed grain.

Note: Shaving the end grain to a smooth surface with a sharp knife can yield interesting results when viewed through a magnifying glass. This can also help you determine if the grain is open or closed. Small pores indicate closed grain, large pores indicate open grain.

Final Project Requirements

- All wood samples
- Completed Wood Properties Lab Sheet

Name _____

Wood Properties Lab Sheet

	Sample 1	Sample 2	Sample 3	Sample 4	Sample 5	Sample 6	Sample 7	Sample 8	Sample 9	Sample 10
	Radially (quarter sawn)						Tangentially (plain sawn)			
Species										
Wet thickness										
Wet width										
Wet length										
Wet weight										
Dry weight										
Moisture Content										
Dry thickness										
Difference wet thickness - dry thickness										
Dry width										
Difference wet width - dry width										
Dry length										
Difference Wet Length - Dry Length										
Open or closed grain										
Color										

Follow-Up Questions

1. What were your main challenges in completing this project?

   _____

   _____

   _____

   _____

2. Is it good or bad for wood to be at zero moisture content before using it in a cabinetmaking project? Explain your answer.

   _____

   _____

   _____

   _____

3. Did thickness, width, or length change the most after drying? Was this as predicted in the text?

   _____

   _____

   _____

4. What species seemed to be the densest and, therefore, the strongest?

   _____

Name _____

Rubric for Moisture Content Material Testing
Student Guidelines

Completeness of lab sheet   Possible points_____   Score_____

Have you completed all sections of the Lab Sheet?

_____

Accuracy of findings   Possible points_____   Score_____

Are your findings comparable to what is expected?

_____

Reporting quality   Possible points_____   Score_____

Do you think you have met the objectives?

_____

Efficiency   Possible points_____   Score_____

Did you finish the work in the allotted time?

_____

Follow-up questions   Possible points_____   Score_____

Were all questions answered in a thoughtful manner?

_____

Total possible points _____   Score_____

Name _____ Date _____ Course _____

SECTION PROJECT 4-1
Making a Bench Hook

When using hand tools, it can be a challenge to hold the stock while it is being worked. Clamps and vises can be inefficient or get in the way of the work. A bench hook or "shooting board" can be used in some of these situations.

Resource Chapters

Chapter 22—Sawing with Hand and Portable Power Tools

Chapter 23—Sawing with Stationary Power Machines

Chapter 24—Surfacing with Hand and Portable Power Tools

Chapter 27—Drilling and Boring

Objectives

After completing this project, you should be able to:

- Use the plans to determine materials needed.
- Use the cabinet workshop to complete a small project.
- Crosscut a small piece of stock to length using a fixture to hold the material.

Materials and Resources

Items you might need to complete this project include:

- 3/4" thick particleboard
- Square
- Table saw
- Wood glue
- Drill press

- 1 1/2" no. 10 screw
- Drill
- 1/8" drill bit
- 3/16" drill bit

SAFETY NOTE
Before proceeding with this activity, you should have passed a general shop safety rule test, all specific safety tests on the tools used in this activity, and be certified by the instructor to use the tools and equipment needed for this activity.

Activity Procedure

1. Study the bench hook plans shown here.

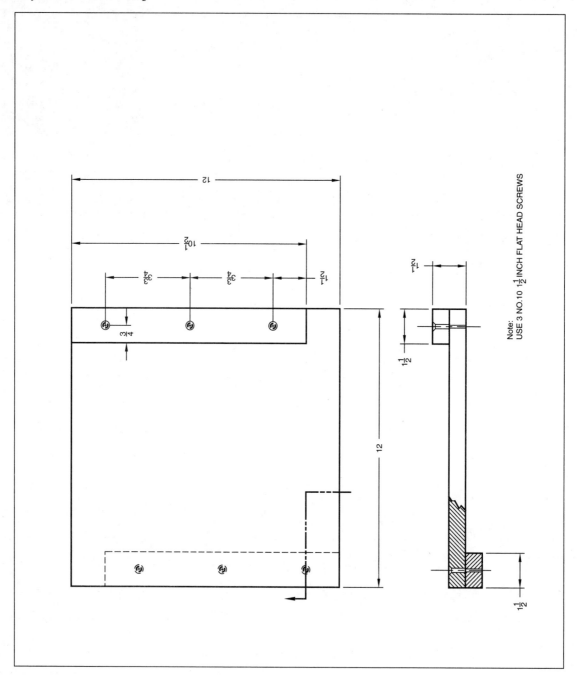

2. Make a cut list.//
3. Make a bill of materials.
4. Make a list of resources needed other than materials such as tools and equipment.
5. Make a plan of procedure. Remember, this is the most important part of getting ready to work. You should be able to follow your plan step-by-step while executing your project.
6. Execute your plan of procedure. Refer to pages 106 and 107 in the text for examples of procedural steps.

Name _____

Final Project Requirements

- Bill of materials
- Plan of procedure
- A completed bench hook
- Follow-up questions

Bill of Materials

Project Name: _____

Quantity	Units	Material Description	Price/Unit	Extended Price
			Subtotal	
			Tax	
			Total	

Material Cut Sheet

Part Name	Material	Number Pieces	Rough Measurements			Cut Process Completed	Finish Measurements			Cut Process Completed
			Thickness	Width	Length		Thickness	Width	Length	

Name _____

Plan of Procedure

Minutes, Hours, Days, Weeks, or Months

Steps

Follow-Up Questions

1. What were your main challenges in completing this project?

   _____
   _____
   _____
   _____

2. Explain how this project will be used.

   _____
   _____
   _____
   _____

3. Did you finish your project on time? What processes went faster than you thought they would and what processes took longer?

   _____
   _____
   _____
   _____

Name _____

Rubric for Making a Bench Hook
Student Guidelines

Materials and resources Possible points_____ Score _____

Did you supply a complete bill of materials, complete list of resources, and total cost of materials calculated?

_____

Plan of procedure Possible points_____ Score _____

Did you create a detailed list of steps to build the project, and keep accurate time records of estimated time and actual time for each step?

_____

Craftsmanship Possible points_____ Score _____

Did you follow plans? Are corners broken properly, edges flush at joints, and screw heads countersunk?

_____

Follow-up questions Possible points_____ Score _____

Were all questions answered in a thoughtful manner?

_____

Total possible points _____ Score _____

Name _____ Date _____ Course _____

SECTION PROJECT 4-2
Drill Press Practice

This activity will provide the opportunity for practice on the drill press.

Resource Chapters

Chapter 22—Sawing with Hand and Portable Power Tools

Chapter 23—Sawing with Stationary Power Machines

Chapter 24—Surfacing with Hand and Portable Power Tools

Chapter 27—Drilling and Boring

Chapter 37—Joinery

Objectives

After completing this project, you should be able to:

- Read a view drawing.
- Set the drill press to a certain depth.
- Change drill bits in a drill press.
- Lay out location of holes.
- Drill holes accurately to a layout.

Materials and Resources

- 2 × 4 pine or similar material
- Fully functional drill press

SAFETY NOTE
Before proceeding with this activity, you should have passed a general shop safety rule test, all specific safety tests on the tools used in this activity, and be certified by the instructor to use the tools and equipment.

Activity Procedure

1. Cut a piece of 2 × 4 material to 6″. A 2 × 4 is 1 1/2″ × 3 1/2″ nominal dimensions.
2. Use a square and pencil to mark the location of the hole on the material.

3. Install the 3/8" twist drill.
4. Set the drill press depth stop so that the 3/8" holes will drill down 3/4".
5. Bore four 3/8" holes in the corners.
6. Install a 3/4" sharp spade bit.
7. Install a drill press vise on the drill press
8. Place a sacrificial board in the vise that is slightly smaller than the workpiece. The sacrificial board will keep the wood from splintering through on the back side and also give the spade bit point something to drill into.
9. Drill the hole.
10. Install a 1/4" drill bit in the drill press.
11. Turn the piece up on edge, reclamp in the vise, and drill the 1/4" hole as per plans.
12. Turn the workpiece up on end, reclamp in the vise, and drill the second 1/4" hole as per plans.
13. Submit your completed project.

Final Project Requirements

- Completed project
- Follow-up questions

Name _____

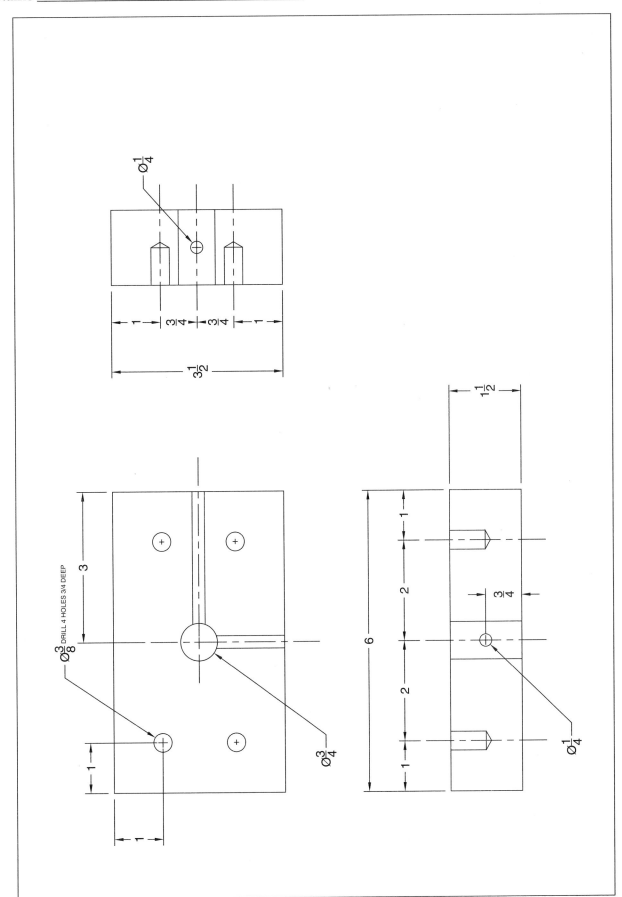

Follow-Up Questions

1. What were your main challenges in completing this project?

   _____
   _____
   _____
   _____

2. Did you finish your project on time? What processes went faster than you thought they would and what processes took longer?

   _____
   _____
   _____
   _____

3. Explain how you kept from drilling the four 3/8" diameter holes too deep.

   _____
   _____
   _____
   _____

4. How did you hold the work while drilling to keep it steady?

   _____
   _____
   _____
   _____

5. What kind of bit did you use to drill the 3/4" hole?

   _____
   _____

Name _____

Rubric for Drill Press Practice
Student Guidelines

Plan of procedure   Possible points_____   Score_____

Did you provide a detailed list of steps to build the project, and keep accurate time records of estimated time and actual time for each step?

_____

Craftsmanship   Possible points_____   Score_____

Did you follow plans, drill holes in proper locations, and follow proper procedure on all operations?

_____

Follow-up questions   Possible points_____   Score_____

Did you answer all questions in a thoughtful manner?

_____

Total possible points _____   Score _____

Name _____ Date _____ Course _____

SECTION PROJECT 4-3
Dry Fit

Each project you make will provide unique challenges in clamping. Most cabinetmakers will dry fit all joints before adding glue. In this step, they will also determine how to apply clamping pressure. Glued joints without pressure do not have the strength of glued joints that have pressure applied.

Resource Chapter

Chapter 32—Gluing and Clamping

Objectives

After completing this project, you should be able to:

- Be able to use all clamps available in your shop to clamp materials in various shapes and sizes.

Materials and Resources

- Variety of bar clamps
- Hand-screw clamps
- C-clamps
- Miscellaneous specialty clamps available in your shop
- Vacuum press
- Variety of wood pieces, cut square and with miters (edges and end should be true, as if they were going to be glued)

Activity Procedure

SAFETY NOTE
Before proceeding with this activity, you should have passed a general shop safety rule test, all specific safety tests on the tools used in this activity, and be certified by the instructor to use the tools and equipment.

1. Find two pieces of wood, approximately 6″ each.
2. Clamp the wood pieces face-to-face using hand-screw clamps. Practice placing the jaws parallel in such a way that pressure is applied evenly across the entire jaw of the clamp. Ask your instructor to observe your setup and mark the rubric section.
3. Find five pieces of wood that are approximately 3/4″ × 3″ × 18″.

4. Using bar clamps, clamp these pieces together as if they are to be joined edge-to-edge, forming a larger piece. Ask your instructor to observe your setup and mark the rubric section.
5. Find two small pieces of wood in the 4"–6" size range.
6. Clamp the wood pieces with C-clamps. The two surfaces should be clamped solidly. Ask your instructor to observe your setup and mark the rubric section.
7. Set up the vacuum bag press. Place two pieces of wood face-to-face and put them in the press. Pull a vacuum on the pieces. Ask your instructor to observe your setup and mark the rubric section.
8. Find a piece of plywood and 1/4" thick edgeband strips.
9. Use edge clamps to dry fit edge trim to the plywood. Ask your instructor to observe your setup and mark the rubric section.

Final Project Requirements

- Follow-up questions

Follow-Up Questions

1. What were your main challenges in completing this activity?

_____
_____
_____
_____

2. Explain how this activity experience will help you with later projects.

_____
_____
_____
_____

3. Explain how you were able to get even pressure with the hand-screw clamps.

_____
_____
_____
_____

Name _____

Rubric for Dry Fit
Student Guidelines

Hand-screw clamps Possible points_____ Score _____

Were jaws clamped parallel with even pressure? Was the correct number of clamps used for the material?

_____

Bar clamps Possible points_____ Score _____

Was the correct number of clamps used for the material? Was the grain alternated?

_____

C-clamps Possible points_____ Score _____

Did you have a detailed list of steps to build the project? Did you keep accurate time records of estimated time and actual time for each step?

_____

Vacuum bag press Possible points_____ Score _____

Was the press set up properly and not leaking?

_____

Edge clamps Possible points_____ Score _____

Were clamps set up properly?

_____

Follow-up questions Possible points_____ Score _____

Did you answer all questions in a thoughtful manner?

_____

Total possible points _____ Score _____

Copyright Goodheart-Willcox Co., Inc.
May not be reproduced or posted to a publicly accessible website.

Name _____ Date _____ Course _____

SECTION PROJECT 4-4
Building a Machine Jig or Fixture

Machine jigs or fixtures can add production efficiency to your work. In this activity, you are going to develop a jig or fixture to make a common procedure more efficient.

Resource Chapters

Chapter 22—Sawing with Hand and Portable Power Tools

Chapter 23—Sawing with Stationary Power Machines

Chapter 24—Surfacing with Hand and Portable Power Tools

Chapter 27—Drilling and Boring

Chapter 37—Joinery

Chapter 38—Accessories, Jigs, and Special Machines

Objectives

After completing this project, you should be able to:

- Explain why shop made accessories, jigs, and fixtures are important for shop production.
- List the most common accessories, jigs, and fixtures.
- Explain how a jig works.

Materials and Resources

- Scrap stock
- Plywood
- MDF
- Screws
- Dowels

Activity Procedure

SAFETY NOTE
Before proceeding with this activity, you should have passed a general shop safety rule test, all specific safety tests on the tools used in this activity, and be certified by the instructor to use the tools and equipment.

1. Decide which jig or fixture you want to make. Trade journals are a good source of ideas. Other ideas include:

    table saw—taper jig, both adjustable and non-adjustable, miter jig, splined plain miter jig, splined flat miter jig, tenon jig, outfeed table

    band saw—circle cutting fixture

    drill press—V-block, dowel pin drill fence

    picture frame clamp

    radial arm saw—push board

    band saw—auxiliary fence

2. Prepare a sketch of the fixture or jig on graph paper.
3. Prepare a plan of procedure.
4. Prepare a bill of materials.
5. Prepare a cut list

Final Project Requirements

- Shop sketch on graph paper
- Bill of materials
- Cut list
- Plan of procedure
- Final fixture
- Demonstration of how your jig or fixture works
- Follow-up questions

Name _____

DRAWING SCALE ____ = ____ SQUARES   DATE __/__/__   FILE No. ____

NAME : ____

Plan of Procedure

Minutes, Hours, Days, Weeks, or Months: 1–25

Steps: 1–32

Name _____

Bill of Materials

Project Name: _____

Quantity	Units	Material Description	Price/Unit	Extended Price
			Subtotal	
			Tax	
			Total	

Material Cut Sheet

Part Name	Material	Number Pieces	Rough Measurements			Cut Process Completed	Finish Measurements			Cut Process Completed
			Thickness	Width	Length		Thickness	Width	Length	

Name _____

Follow-Up Questions

1. What were your main challenges in completing this project?

   _____
   _____
   _____
   _____

2. Explain how this project will be used.

   _____
   _____
   _____
   _____

3. What accessory, jig, or fixture did you construct?

   _____
   _____
   _____
   _____

4. Explain the difference between a jig and a fixture.

   _____
   _____
   _____
   _____

Rubric for Building a Machine Jig or Fixture
Student Guidelines

Print Possible points _____ Score _____

Assess your project neatness, accuracy, dimensions, and correctness of views.

_____

Plan of procedure Possible points _____ Score _____

Did you have a detailed list of steps to build the project, and keep accurate time records of estimated time and actual time for each step?

_____

Demonstration Possible points _____ Score _____

Were you confident and organized while demonstrating your project?

_____

Craftsmanship Possible points _____ Score _____

Did you follow plans? Did joint represent perfect location of holes and cuts?

_____

Follow-up questions Possible points _____ Score _____

Did you answer all questions in a thoughtful manner?

_____

Total possible points _____ Score _____

Name _____ Date _____ Course _____

SECTION PROJECT **4-5**
Wood Joints Class Project

A thorough knowledge of joinery techniques is required to be considered a master cabinetmaker. In this activity, each student will be assigned one or two joints shown in the textbook chapter and make a sample of each.

Resource Chapters

Chapter 22—Sawing with Hand and Portable Power Tools

Chapter 23—Sawing with Stationary Power Machines

Chapter 24—Surfacing with Hand and Portable Power Tools

Chapter 27—Drilling and Boring

Chapter 37—Joinery

Objectives

After completing this project, you should be able to:

- Explain the technique to make at least one wood joint.
- Identify the joints in the activity.
- Identify the difference in non-positioned, positioned, and reinforced joints.

Materials and Resources

- Scrap wood
- Assorted finish nails
- Assorted screws
- 3/8" dowels
- No. 20 biscuits
- Pocket screws

- Plate joiner
- Dowel jig
- Tenon jig
- Pocket screw jig or machine
- 4' × 4' board for mounting joints

SAFETY NOTE
Before proceeding with this activity, you should have passed a general shop safety rule test, all specific safety tests on the tools used in this activity, and be certified by the instructor to use the tools and equipment needed for this activity.

Copyright Goodheart-Willcox Co., Inc.
May not be reproduced or posted to a publicly accessible website.

Activity Procedure

1. With your classmates' help, cut the scrap wood into 3″ × 6″ pieces.
2. Collect two pieces of wood to build your joint. Joints to consider include:

    Box joint

    Dovetail joint

        Handmade through single

        Dovetail joint, machine-made through multiple

        Dovetail joint, half-blind multiple

    Face-to-face butt joint

    Edge-to-edge butt joint

    End-to-end butt joint

        Reinforced with dowels

    End-to-edge butt joint

        Reinforced with pocket screws

    End-to-face butt joint

        Reinforced with screws

        Reinforced with nails

        Reinforced with dowels

        Reinforced with glue block

        Reinforced with RTA fasteners

    Dado joint

        Blind dado

        Half dado

    Groove joint

    Rabbet joint

        Half rabbet

        Dado and rabbet

        Dado tongue and rabbet

    Lap joints

        T-lap

        End lap

        Middle lap

    Miter joints

        Flat miter

        Plain miter

        Rabbet miter

        Half lap miter

        Lap and miter

    Mortise and tenon

        Blind mortise and tenon

        Through mortise and tenon

        Open mortise and through tenon

    Butterfly joint

3. Refer to the text to learn how to make your joint.
4. Sketch your joint on graph paper.
5. Make a plan of procedure.
6. Make your joint.
7. Prepare a five-minute oral report about the joint you made.

Final Project Requirements

- Sample assigned joint
- Report outline
- Follow-up questions

Name _____

DRAWING SCALE _____ = _____ SQUARES   DATE ___/___/___   FILE No. _____

NAME : _____

Plan of Procedure

Steps | **Minutes, Hours, Days, Weeks, or Months** (1–25)

(Blank chart with steps 1–32)

Name _____

Follow-Up Questions

1. What were your main challenges in completing this project?

   _____

   _____

   _____

   _____

2. Explain how this project will be used.

   _____

   _____

   _____

   _____

3. What joints did you complete?

   _____

   _____

   _____

   _____

Rubric for Wood Joints Class Project
Student Guidelines

Print Possible points_____ Score _____

Assess neatness, accuracy, dimensions, correctness of views, and accuracy of plan of procedure.

_____

Plan of procedure Possible points_____ Score _____

Did you have a detailed list of steps to build the project, and keep accurate time records of estimated time and actual time for each step?

_____

Oral presentation Possible points_____ Score _____

Were you confident and organized in presenting information?

_____

Craftsmanship Possible points_____ Score _____

Did you follow plans? Did joints represent perfect location of holes and cuts?

_____

Follow-up questions Possible points_____ Score _____

Were all questions answered in a thoughtful manner?

_____

Total possible points _____ Score _____

Name _____ Date _____ Course _____

SECTION PROJECT **4-6**

Make a Push Stick from a Pattern

Using push sticks to keep fingers a safe distance from machinery saw blades is a habit all cabinetmakers should create. In this activity, you will make several push sticks using a pattern. It is a good idea to have multiple push sticks on hand because one is used as a sacrificial stick when cutting material narrower than 3/4" wide.

Resource Chapters

Chapter 12—Measuring, Marking, and Laying Out Materials

Chapter 22—Sawing with Hand and Portable Power Tools

Chapter 23—Sawing with Stationary Power Machines

Objectives

After completing this project, you should be able to:

- Transfer a design from a square grid layout to a pattern.
- Use the band saw or jig saw to cut out an irregular shape.

Materials and Resources

- 8 1/2" × 11" card stock paper
- 8 1/2" × 11" plywood or solid wood
- 1/8" or 1/4" tempered hardboard for an original pattern
- Portable jig saw and band saw

SAFETY NOTE
Before proceeding with this activity, you should have passed a general shop safety rule test, all specific safety tests on the tools used in this activity, and be certified by the instructor to use the tools and equipment needed for this activity.

Activity Procedure

1. Find a piece of 8 1/2" × 11" card stock and mark off a 1/2" grid pattern.
2. Using the activity sheet provided, lay out the pattern as explained in Chapter 12.
3. Glue this pattern to the piece of hardboard.
4. Using the band saw, cut out the pattern.
5. Using a disk sander, sandpaper, or files, smooth the edges, making them true to the pattern drawing.

6. Use the pattern to lay out two more push sticks. Retain this pattern for future use.
7. Cut one push stick using the band saw and following the principles explained in Chapter 23.
8. Cut another push stick using the jig saw and following the principles explained in Chapter 22.

Final Project Requirements

- Hardboard pattern
- Two 3/4" plywood or solid wood push sticks as per drawing

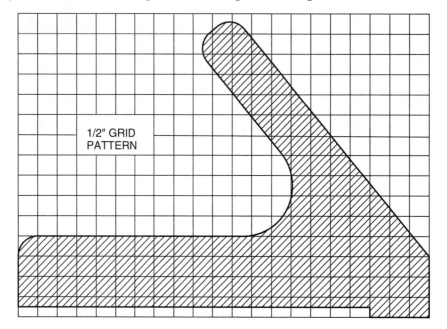

Follow-Up Questions

1. What were your main challenges in completing this project?

_____

_____

_____

2. Explain how this project will be used.

_____

_____

_____

3. Explain the procedure you used to transfer the design from the print to your pattern.

_____

_____

_____

Name _____

Rubric for Make a Push Stick from a Pattern
Student Guidelines

Pattern layout Possible points_____ Score _____

Did you follow the procedure in laying out grid and pattern?

_____

Materials and Resources Possible points_____ Score _____

Did you supply a complete bill of materials, complete list of resources, and determine total cost of materials?

_____

Craftsmanship Possible points_____ Score _____

Did you follow plans, sand out mill marks, round edges, and finish?

_____

Follow-up questions Possible points_____ Score _____

Were all questions answered in a thoughtful manner?

_____

Total possible points _____ Score _____

Name _____ Date _____ Course _____

SECTION PROJECT **4-7**
Make a Routed Sign

There is an entire industry dedicated to making wood signs. Many communities have restrictions on lighted signs, creating an opportunity for cabinetmakers to create wood signs. Routers are typically used to make wood signs. This project will allow you to get familiar with using a router.

Resource Chapters

Chapter 10—Sketches, Mock-Ups, and Working Drawings

Chapter 12—Measuring, Marking, and Laying Out Materials

Chapter 26—Shaping

Chapter 29—Abrasives

Objectives

After completing this project, you should be able to:

- Properly select and change router bits.
- Demonstrate mastery of edge routing.
- Demonstrate mastery of freehand control of the router to follow a line.

Materials and Resources

- 3/4″ × 5″ × 16″ solid stock MDF or solid wood stock
- Router
- V-groove bit for routing words or numerals
- Roman ogee or 3/8″ round-over for adding a decorative edge

SAFETY NOTE
Before proceeding with this activity, you should have passed a general shop safety rule test, all specific safety tests on the tools used in this activity, and be certified by the instructor to use the tools and equipment needed for this activity.

Activity Procedure

1. Use graph paper to design a sign on 5″ × 16″ stock.
2. Lay out the full-size sign on white paper.

3. Glue the paper to the wood with spray adhesive. Don't use too much adhesive, or you will have trouble peeling off the paper.

4. Install the V-groove bit into the router and adjust to proper depth. Test this on scrap wood first.

5. Secure sufficient lighting and fasten material properly to the table. Put on hearing protection.

6. Turn on the router and carefully place it on your line.

7. Steadily follow the line and lift up the router when you reach the end.

    Tip: Allow the edges of your hand to rest on the material and control the router with your fingers. This will give you more control of the router. Make sure to angle your sight so you can see your line as you proceed with the cut.

8. When finished with the routing, change the router bit to the round-over or roman ogee bit.

9. Adjust the bit depth to get the desired edge profile. Test this on scrap wood first.

10. Turn the router on and proceed counterclockwise around the sign. You will be forcing the router as far as it will go into the piece. The bearing will keep you from cutting too deep.

11. When finished, check the edge to make sure you followed the wood edge all the way around. If you let the router come away, go back over it.

    Tip: Depending on how much the router cuts, you may want to make a pass with the router at less-than-final depth to avoid splintering the wood and overloading the router. Some wood splinters more than others. Work carefully when finishing on the ends.

12. Peel the paper off and sand. If you used too much spray adhesive, you may have to sand the paper off.

13. Apply desired finish.

Final Project Requirements

- Shop sketch
- Plan of procedure
- Bill of materials
- Final sign project
- Follow-up questions
- Student-produced activity print

Name _____

DRAWING SCALE _____ = _____ SQUARES   DATE ___/___/___   FILE No. _____

NAME : _____

Plan of Procedure

Name _____

Bill of Materials

Project Name: _____

Quantity	Units	Material Description	Price/Unit	Extended Price
			Subtotal	
			Tax	
			Total	

Copyright Goodheart-Willcox Co., Inc.
May not be reproduced or posted to a publicly accessible website.

Follow-Up Questions

1. What were your main challenges in completing this project?

   _____

   _____

   _____

   _____

2. Explain how this project will be used.

   _____

   _____

   _____

   _____

3. Did you finish your project on time? What processes went faster than you thought they would and what processes took longer?

   _____

   _____

   _____

   _____

4. If you made $10.00 per hour, what would you get paid for this project?

   _____

   _____

5. Did you have any trouble keeping the router on line when you were routing your letters?

   _____

   _____

6. Explain how you kept the router steady when working freehand.

   _____

   _____

   _____

   _____

Name _____

Rubric for Make a Routed Sign
Student Guidelines

Shop sketch Possible points _____ Score _____
Assess your neatness, accuracy, dimensions, and correctness of views.

_____

Materials and resources Possible points _____ Score _____
Did you supply a complete bill of materials, complete list of resources, and total cost of materials?

_____

Router Possible points _____ Score _____
Did you change bit without assistance? Is letter engraving steady, edge shaping done properly?

_____

Craftsmanship Possible points _____ Score _____
Were plans followed, mill marks sanded out, corners broken properly, edges flush at joints, and screw heads countersunk?

_____

Follow-up questions Possible points _____ Score _____
Were all questions answered in a thoughtful manner?

_____

Total possible points _____ Score _____

Name _____ Date _____ Course _____

SECTION PROJECT **4-8**
Make a Sanding Block

Properly finished workpieces usually involve some hand sanding. This step should be done carefully. Incorrect sanding in the final stages of project completion can make your work look substandard. There are many commercial sanding blocks on the market, but most cabinetmakers make their own. In this activity, you will learn how to tear sandpaper and make a sanding block.

Resource Chapter
Chapter 29—Abrasives

Objectives
After completing this project, you should be able to:
- Explain several methods for tearing sandpaper.
- Explain why using three layers of sandpaper is the best technique.
- Explain the abrading process.

Materials and Resources
- 3/4" × 2 1/4" × 5 1/2" MDF or any suitable scrap wood that can be sawn
- One sheet of sandpaper, any grit

SAFETY NOTE
Before proceeding with this activity, you should have passed a general shop safety rule test, all specific safety tests on the tools used in this activity, and be certified by the instructor to use the tools and equipment needed for this activity.

Activity Procedure

1. Cut the MDF into 3/4″ × 2 1/4″ × 5 1/2″ dimensions.

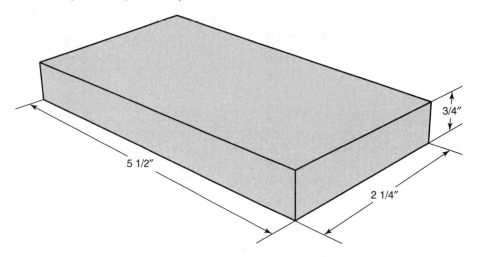

2. Fold a piece of 80 or 100 grit paper as shown in the following diagram. Make the fold so that the abrasive is on the outside of the fold and the plain paper is inside the fold.

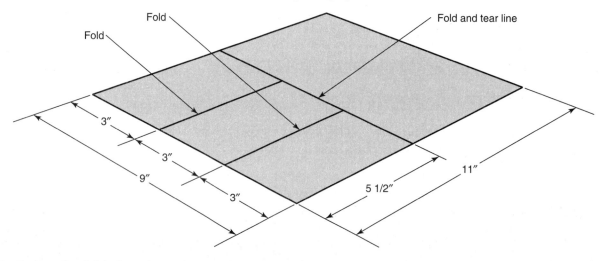

3. Crease the fold sharply with your fingernail or a dowel. The paper will almost tear itself at this point. If you have trouble tearing the paper, make a small tear in the fold to start the tear. Then place the fold over the edge of a table and finish the tear. Now you have half a sheet.

4. Divide this sheet into thirds and fold as the diagram above indicates. This creates the perfect size of paper for the sanding block you made in Step 1.

Note: Another benefit to folding the sandpaper into thirds is that it can become a gripping tool. Place the folded paper between the workbench and a workpiece you are clamping to the bench. It will make the workpiece nearly immovable. Sandpaper folded into thirds for hand sanding is more stable in your hands. Compare with a piece of sandpaper that is less than one-half sheet. When you do not fold at all, you, your block, or hand will slide on the paper. When you double the sheet, it will slide between the two halves. But when you fold in thirds, it grips on both sides and the sheets will not slide between the folds. Experiment with this explanation and report your findings in the follow-up questions.

Name _____

Final Project Requirements
- Sanding block
- Properly folded sandpaper
- Follow-up questions

Follow-Up Questions

1. What were your main challenges in completing this activity?

   _____

   _____

   _____

   _____

2. Explain how this project will be useful to you when building cabinets.

   _____

   _____

   _____

   _____

3. Did you have more control of the sandpaper when it was folded into thirds? Explain.

   _____

   _____

   _____

   _____

Rubric for Make a Sanding Block
Student Guidelines

Sandpaper block   Possible points _____   Score _____

Were measurements accurate and according to plans?

_____

Folding and tearing sandpaper   Possible points _____   Score _____

Was paper folded and torn according to plans?

_____

Follow-up questions   Possible points _____   Score _____

Were all questions answered in a thoughtful manner?

_____

                                       Total possible points _____   Score _____

Name _____ Date _____ Course _____

SECTION PROJECT **4-9**
Build a Sawhorse

In this project, you will build a sawhorse out of 2 × 4 lumber, using a portable circular saw and wide-blade handsaw. The challenge of this project is to keep the angles from becoming confusing. Work systematically to avoid that issue. Using a compound miter saw also helps simplify the process. The resulting sawhorse is strong, if the craftsmanship is good. A gusset or cross support can be added to make it even stronger.

Resource Chapters

Chapter 22—Sawing with Hand and Portable Power Tools

Chapter 23—Sawing with Stationary Power Machines

Objectives

After completing this project, you should be able to:

- Cut dimension lumber to length using a standard wide-blade crosscut handsaw. Pieces will be cut the correct length and perfectly square (90°) to the edges.
- Use a portable circular saw to cut a compound miter on dimension lumber. Cuts will be the specified angles when finished.
- Use a wide-blade handsaw or backsaw to make specialty cuts in dimension lumber.
- Properly set up the workstation for portable power equipment use.
- Determine how to clamp and support material being cut with portable power equipment and hand tools.

Materials and Resources

- Two 2" × 4" × 96" studs (actual size 1 1/2" × 3 1/2" × 96")
- Eleven 2" wood screws
- Variable speed drill
- 3/16" drill bit
- 3/8" inch drill bit
- Phillips bit
- Wide-blade handsaw
- Portable circular saw
- Sliding T-bevel
- Square that can be used to lay out angles
- Pencil
- C-clamps

SAFETY NOTE

Before proceeding with this activity, you should have passed a general shop safety rule test, all specific safety tests on the tools used in this activity, and be certified by the instructor to use the tools and equipment needed for this activity.

Activity Procedure

1. Prepare workstation by running extension cords, gathering material, and planning how to support the material.

2. Set the angle on the portable circular saw to 15°.

3. Using the angle square, lay out a 10° angle on the end of the 2 × 4. Start the line approximately 1/2″ from the end to allow for some drop-off material. Be sure the saw kerf is to the right side of the line, the fall-off side of the material.

   Note: It is usually easier to obtain a true cut with a portable circular saw if you are not cutting right on the end of material.

4. Make a cut on the 10° layout line with the saw set to 15°. This will give you a compound miter cut. This is a cut that yields two angles on a piece of stock.

5. Measure 24″ from the longest point of the previous cut. Mark and lay out another 10° angle that is parallel to the first cut and make the cut. This time the saw kerf should be on the left side of the layout line. Make the cut. The leg piece you are cutting is actually the fall-off material.

SAFETY NOTE

Make sure you are not cutting between two supports as this can cause a kickback. As the cut is being made, the material will sag and pinch the saw. The leg should fall off. Have someone support it slightly as the cut is finished so the piece will not break off.

6. If your last cut went smoothly, you should have the angle for the next leg already cut on the end of the remaining stock. If it did not cut smoothly, then simply move in about 1/2″ and lay out a 10° line as you did at first and make another cut. Then repeat step 5. This will yield the second leg piece.

7. On the remaining stock, mark a 90° line with your square across the end of the piece, approximately 1/2″ from the end and cut this piece.

8. Measure 36″ from the end you just cut, mark another 90° line with your square, and make the cut.

   Note: For the next cut, you will want to rotate the 36″ piece so the cut is still on your right side. Otherwise, you would have to move around to the other side of the piece to make the cut, in order to keep the largest part of saw base plate supported.

9. Repeat steps 3 to 8 on the second 2″ × 4″ × 96″ piece of material. You now have four legs cut and the two horizontal 36″ pieces for the sawhorse.

10. What remains is to cut the cheek cuts on the legs. Orient the legs in the position they will be in when assembled and mark the ends and side that will mate up to the horizontal piece. It is easy to confuse yourself on this and cut the wrong end, so lay it out on the floor to get the cuts straight in your mind.

11. Draw lay out lines for the cheek cuts. Refer to the diagram.

12. Using a wide-blade handsaw, make the cheek cuts. There are various ways to support your work for this process. Be sure, however, to set up the material so that it will not move while cutting and so that you can get a full, comfortable stroke on the saw.

13. Drill three evenly spaced 3/16″ holes in the face of one of the 36″ 2 × 4s.

Name _____

14. Counterbore the previous holes one inch deep with the 3/8″ bit.

15. Attach the two 36″ pieces together with 2″ wood screws.

16. Drill two 3/16″ pilot holes about 2″ from the top of each leg piece and 3/4″ from the edge. This will keep the wood from splitting when the screw is driven. Cut the legs.

17. Use the variable speed drill to drive the screws to attach the legs to the horizontal piece. Refer again to the diagram.

Final Project Requirements

- Completed sawhorse
- Follow-up questions

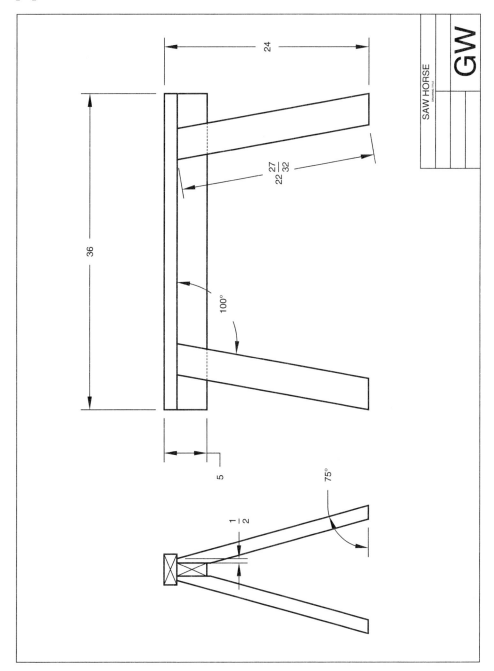

Follow-Up Questions

1. What were your main challenges in completing this activity?

   _____
   _____
   _____
   _____

2. What is a saw kerf?

   _____
   _____
   _____
   _____

3. Explain how to make a compound angle.

   _____
   _____
   _____
   _____

Name _____

Rubric for Build a Sawhorse
Student Guidelines

Measurements　　　　　　　Possible points _____　Score _____

Were measurements accurate according to plans?

_____

Assembly　　　　　　　　　Possible points _____　Score _____

Were measurements accurate according to plans?

_____

Cuts　　　　　　　　　　　Possible points _____　Score _____

Were cuts clean, accurate, and at proper angles?

_____

Follow-up questions　　　　Possible points _____　Score _____

Were all questions answered in a thoughtful manner?

_____

　　　　　　　　　　　　　　　Total possible points _____　Score _____

Name _____ Date _____ Course _____

SECTION PROJECT **4-10**
Sharpening a Chisel

Cabinetmakers must know how to maintain a keen edge on a chisel because dull chisels are useless and dangerous. In this activity, you will practice sharpening a chisel. A properly sharpened chisel is razor sharp.

Resource Chapters

Chapter 24—Surfacing with Hand and Portable Power Tools

Chapter 39—Sharpening

Objectives

After completing this project, you should be able to:

- Explain the concept of hollow ground.
- Determine when a chisel needs only honing or needs both grinding and honing.
- True the end of a chisel.
- Properly grind a chisel.
- Properly hone a chisel.
- Explain the safety principles to follow when sharpening and working with chisels.

Materials and Resources

- Grinder with medium and fine stone and adjustable tool supports
- Dull chisel
- Try square
- Cup of cooling water
- Sharpening stones, 800 to 6000 grits

SAFETY NOTE
Before proceeding with this activity, you should have passed a general shop safety rule test, all specific safety tests on the tools used in this activity, and be certified by the instructor to use the tools and equipment needed for this activity.

Activity Procedure

1. Make a profile drawing of how you will sharpen the chisel. Make your drawing similar to **Figure 24-5** in the text. The drawing is a plane iron, not a chisel, but the profile is the same.

2. Write a plan of procedure.

3. Determine if chisel needs grinding and honing or honing only.

4. Check the end of the chisel with a square to ensure the end is perfectly square.
5. Use the sharpening stone with the coarsest grit to true the end. Hold the chisel at a 90° angle to the stone and hone in a figure eight pattern.
6. If the edge is far from square, you may have to grind the end square. If you have to do this, quite a bit of grinding will be required to get the hollow ground edge back.
7. Once the end is square, set the tool rest so that the grinding wheel will grind a 25° angle on the end of the chisel.
8. For safety, the tool rest should not be more than 1/8" inch away from the grinding wheel.
9. Move the chisel carefully back and forth, making sure to keep the grinding even. Watch the tip carefully. Do not allow it to overheat. Dip the edge frequently into the water to keep the edge cool.
10. Grind the chisel until the ground surface meets the tip and you can no longer see a shine on the tip. Look directly at the end with light behind you.
11. Now you are ready to start the honing process. You must keep the chisel at a 30° angle while honing. Use a honing fixture, if available. Otherwise, get a feel for the angle and maintain that angle while honing. Honing by feel takes practice. You must maintain the same angle or you will not get satisfactory results.
12. Test the chisel sharpness by following some of the suggested tests discussed in the textbook.

Final Project Requirements

- Profile drawing
- Plan of procedure
- Sharp chisel
- Follow-up questions

Follow-Up Questions

1. What were your main challenges in completing this activity?

   _____

   _____

   _____

2. Why is it important to have a sharp chisel?

   _____

   _____

   _____

3. Did you have all the equipment needed to perform the task?

   _____

   _____

   _____

Name _____

Rubric for Sharpening a Chisel
Student Guidelines

Profile drawing   Possible points _____   Score _____

Assess neatness, accuracy, dimensions, and correctness of views.

_____

Plan of procedure   Possible points _____   Score _____

Was there a detailed list of steps to build the project? Did student keep accurate time records of estimated time and actual time for each step?

_____

Craftsmanship   Possible points _____   Score _____

Did you use proper grinding and honing techniques?

_____

Follow-up questions   Possible points _____   Score _____

Were all questions answered in a thoughtful manner?

_____

                                          Total possible points _____   Score _____

Copyright Goodheart-Willcox Co., Inc.
May not be reproduced or posted to a publicly accessible website.

Name _____ Date _____ Course _____

SECTION PROJECT **4-11**

Make a Sign with a CNC Router

A CNC router can be used to do just about any operation that can be done with traditional hand tools and portable power and stationary power machines. The CNC router has revolutionized the making of wood signs. What used to take hours can now be done in minutes. Have fun designing and making your sign.

Resource Chapters

Chapter 11—Creating Working Drawings

Chapter 28—Computer Numerically Controlled (CNC) Machinery

Objectives

After completing this project, you should be able to:

- Make a simple rectangular shape on a computer using CAD software.
- Export a DXF file from CAD software to CAD/CAM software.
- Generate the tool paths and machine code file in the CAD/CAM software.
- Set up and run the CNC machine to make a simple sign.

Materials and Resources

- CNC router
- CAD software
- 3/4" MDF blank size of your design

SAFETY NOTE
Before proceeding with this activity, you should have passed a general shop safety rule test, all specific safety tests on the tools used in this activity, and be certified by the instructor to use the tools and equipment needed for this activity.

Activity Procedure

1. Sketch out ideas on a piece of graph or plain paper.
2. Draw the design using CAD software.
3. Make a DXF file of the drawing.

Copyright Goodheart-Willcox Co., Inc.
May not be reproduced or posted to a publicly accessible website.

4. Import the file into the CAD/CAM software and design the tool paths.//
5. Export the file to the machine controller.
6. Fixture the wood blank onto the bed of the router.
7. Zero the router.
8. Start the program and cut the sign.

Final Project Requirements

- Preliminary scale sketches on paper
- CAD drawing of your design
- Machine code file
- Final product made on the CNC router
- Follow-up questions

Follow-Up Questions

1. What were your main challenges in completing this project?

_____

_____

_____

_____

2. Explain how the finished product will be used.

_____

_____

_____

_____

3. How many hours did you spend on this project?

_____

4. Why does the machine have to be zeroed in?

_____

_____

_____

_____

Name _____

Rubric for Make a Sign with a CNC Router
Student Guidelines

Sketch Possible points _____ Score _____

Did you create a detailed, accurate description and graphic representation of the final sign?

_____

CAD drawing Possible points _____ Score _____

Did you successfully complete the drawing? Was time on computer spent efficiently? Did time spent demonstrate understanding?

_____

Tool paths Possible points _____ Score _____

Were tool paths efficient and practical? Did time spent demonstrate understanding?

_____

Program execution Possible points _____ Score _____

Was part mounted properly in the machine? Was machine zeroed in properly the first time?

_____

Follow-up questions Possible points _____ Score _____

Were all questions answered in a thoughtful manner?

_____

Total possible points _____ Score _____

Name _____ Date _____ Course _____

SECTION PROJECT **4-12**
Veneering

The idea behind veneering is to make a less expensive wood look and feel like a more expensive wood. In this activity, you will practice veneering techniques on a 6″ × 6″ × 3/4″ MDF core.

Resource Chapters

Chapter 22—Sawing with Hand and Portable Power Tools

Chapter 23—Sawing with Stationary Power Machines

Chapter 30—Using Abrasives and Sanding Machines

Chapter 31—Adhesives

Chapter 32—Gluing and Clamping

Chapter 34—Overlaying and Inlaying Veneer

Objectives

After completing this project, you should be able to:

- Understand the veneering process.
- Cut veneer pieces and join them with veneer tape.

Materials and Resources

- Veneer press or vacuum bag press
- 3/4″ × 6″ × 6″ MDF
- 6″ × 6″ backer board
- Veneer pieces of different kinds of wood
- Veneer tape or high-quality masking tape
- Edgebanding
- Edgebanding machine or simple veneer iron or household electric iron
- Veneer saw
- Hand plane
- Veneer glue
- Wax paper

Copyright Goodheart-Willcox Co., Inc.
May not be reproduced or posted to a publicly accessible website.

485

SAFETY NOTE

Before proceeding with this activity, you should have passed a general shop safety rule test, all specific safety tests on the tools used in this activity, and be certified by the instructor to use the tools and equipment.

Activity Procedure

1. Cut veneer with a veneer saw or utility knife to a rough size of 3 1/8″ × 3 1/8″.
2. Make a trimming clamp as shown in **Figure 34-7A** of the text.
3. Plane each edge to make the pieces exactly 3″ × 3″. Make sure the pieces are perfectly square. Glue sandpaper on the surfaces of your trimming clamp to ensure the veneer does not slip. Make a 3″ × 3″ block to use to mark the finished size of the veneer.
4. Lay out the veneer on the 6″ × 6″ MDF in one of the patterns shown in **Figure 34-5** of the text. Make sure the veneer joints are perfect with no gaps.
5. Tape the joints with veneer tape as shown in **Figure 34-8** of the text.
6. Set up the veneer press or the vacuum bag by getting the cauls ready. Wax paper should separate the cauls from the pieces to be clamped. Otherwise, you might glue your project to the cauls.
7. Apply glue to the backer board and attach to the 6″ × 6″ piece, and then apply the glue to the surface of the MDF and attach the veneer assembly.
8. Set the "sandwich" into the press or vacuum bag between the wax paper and the cauls.
9. Apply pressure or the vacuum.
10. Let set in the press at least two hours.
11. Let the assembly dry 24 hours before working.
12. Clean glue with scrapers, and then sand. Be careful, if scraping, to not damage veneer surface. If there is a lot of excess glue, note that you have used too much glue and can use less next time.
13. Carefully prepare the edges for the edge trim. This can be done on a edge sander or with a hand plane. A setup similar to the one shown in **Figure 30-6** of the text can be used to sand the edges, be careful because you don't want to round the edges.
14. Use veneer edge trim with hot-melt glue coating to edge trim the four edges of the 6″ × 6″ piece. One option is to make 1/16″ edge trim and glue with contact adhesive.
15. Sand the edges to bring the edge trim flush with the surfaces.

Final Project Requirements

- 6″ × 6″ piece
- Follow-up questions

Follow-Up Questions

1. What were your main challenges in completing this activity?

_____

_____

_____

Name _____

2. Explain how the skills developed in the activity can be used to build cabinets.

_____

_____

_____

3. What veneer placement pattern did you use to order your four veneer pieces on the MDF? See **Figure 34-5** in the text. Explain your choice.

_____

_____

_____

Rubric for Veneering
Student Guidelines

Veneering Possible points _____ Score _____

Are veneer joints tight?

_____

Veneer pattern Possible points _____ Score _____

Did you lay out squares in an attractive manner?

_____

Edgebanding Possible points _____ Score _____

Is edgebanding solidly glued to the edges, and are corners trimmed correctly?

_____

Sanding and finish Possible points _____ Score _____

Are imperfections puttied using good technique? Are edges and corners sanded well?

_____

Follow-up questions Possible points _____ Score _____

Were all questions answered in a thoughtful manner?

_____

Total possible points _____ Score _____

Name _____ Date _____ Course _____

SECTION PROJECT **4-13**
Edge Trim

Manmade materials provide many advantages to production cabinetmaking. A disadvantage, however, is that the edges of the material are unattractive in most cases. Therefore, it is necessary to "dress up" the edges to hide the unsightly edges of plywood, MDF, and melamine.

Resource Chapters

Chapter 16—Manufactured Panel Products

Chapter 22—Sawing with Hand and Portable Power Tools

Chapter 23—Sawing with Stationary Power Machines

Objectives

After completing this project, you should be able to:

- Apply an edgeband that has hot-melt glue coating.
- Be able to apply a regular veneer edgeband.
- Use shop equipment to cut material to specified sizes.

Materials and Resources

- 4" × 5" maple plywood
- 1/4" × 13/16" × 4" solid wood maple edgeband
- Maple edgeband with hot-melt glue coating
- Veneer wood edgeband (yellow glue applied and clamped)
- 3/4" × 1 1/4" × 4" solid maple
- Small quick clamps and glue cauls
- Veneer iron or regular household iron
- Wax paper

SAFETY NOTE
Before proceeding with this activity, you should have passed a general shop safety rule test, all specific safety tests on the tools used in this activity, and be certified by the instructor to use the tools and equipment.

Activity Procedure

1. Cut the 4″ × 5″ plywood.
2. Make the 1/4″ × 13/16″ × 4″ solid wood edgeband.
3. Make the 3/4″ × 1 1/4″ solid wood edge trim.
4. Cut a 5 1/8″ piece of the edge trim with the hot-melt glue coating off the roll.
5. Cut a 5 1/8″ piece of the veneer edgeband off the roll.
6. Attach the edge trim with the hot-melt glue coating to a 5″ edge. Make sure to center the banding so that a little sticks past both surfaces. Heat with an iron and press on it with a J-roller or a wooden press-down stick until it cools.
7. The edges that stick past the surfaces must be edge trimmed with a router and carefully sanded. If care is taken, you will hide the edge of the plywood so well that it will be hard to tell it is not solid wood.
8. The next step requires glue, clamps, and wax paper strips. Apply a thin but solid film of glue to the other 5″ edge of the plywood and veneer edgeband.
9. Press the two parts together. The quick setting yellow glue will begin to set the edgeband and hold it in place. Place wax paper strips on the edgeband and then the caul. Clamp the assembly with two clamps. After about an hour, you can remove the clamps. It will be 2–3 hours before you can actually work on the piece again. The edges will be edge trimmed with a router and sanded as was done on the edge with the hot-melt glue coating.
10. Next, cut the groove in the edge of the plywood as the plans indicate.
11. Cut the 1 1/4″ × 4″ piece out of solid maple stock.
12. Cut the tongue on the 1 1/4″ × 4″ piece as plans indicate. (This step would be more easily done in a larger piece and then cut to length and distributed). You may want to work with your other classmates to accomplish this.
13. Cut the 1/4″ × 13/16″ piece out of solid maple stock.
14. Apply glue to the 1/4″ × 4″ piece and the 1 1/4″ × 4″ piece and the edges of the plywood. You are going to clamp this together at the same time. Place wax paper between the caul and the 1/4″ piece and clamp with the two clamps. You will not need a caul for the 1 1/4″ piece.
15. Let the assembly dry a minimum of two hours and sand the edges flush with the face. The 1/4″ piece may need to be edge trimmed and then sanded.

Final Project Requirements

- The piece completed as the plans indicate
- Follow-up questions

Name _____

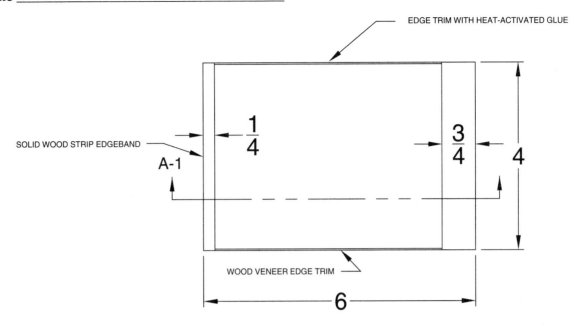

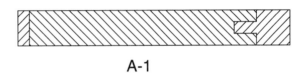

A-1

Follow-Up Questions

1. What were your main challenges in completing this project?

_____

_____

_____

2. Explain which edgebanding technique you liked most, and explain why.

_____

_____

_____

Rubric for Edge Trim
Student Guidelines

All wood edge trim Possible points _____ Score _____

Are the tongue and groove tight, sanded flush with surface, with no glue showing?

_____

1/4" edge trim Possible points _____ Score _____

Did you ensure there are no hollow spots, that trim is sanded flush with face, no glue is showing, and there is solid mating of band to plywood?

_____

Veneer edge trim Possible points _____ Score _____

Did you ensure there are no hollow spots, corners are sanded well, and no glue is showing?

_____

Edge trim with hot-melt glue coating Possible points _____ Score _____

Did you ensure there are no hollow spots, corners are sanded well, and no glue is showing?

_____

Craftsmanship Possible points _____ Score _____

Did you follow plans, sand out mill marks, and ensure corners are broken properly?

_____

Follow-up questions Possible points _____ Score _____

Were all questions answered in a thoughtful manner?

_____

Total possible points _____ Score _____

Name _____ Date _____ Course _____

SECTION PROJECT 4-14
Practice Applying Plastic Laminate

In this project, you will follow the plans to make a simulated kitchen countertop out of particleboard. Plastic laminate will be applied to the surface and front edge as per plans.

Resource Chapters
Chapter 17—Veneers and Plastic Overlays
Chapter 33—Bending and Laminating

Objectives
After completing this project, you should be able to:
- Apply plastic laminate to panel using contact cement.
- Apply plastic laminate to an edge.
- Apply heat-activated laminate.

Materials and Resources
- 4″ × 6″ industrial particleboard
- 1 1/2″ × 6″ industrial particleboard
- 4 1/4″ × 6 1/2″ plastic laminate
- 1 3/4″ × 6″ plastic laminate
- Contact cement (water-based cement if ventilation is suspect)
- Disposable 1″ brushes
- Nail gun
- J-roller
- Yellow glue

SAFETY NOTE
Before proceeding with this activity, you should have passed a general shop safety rule test, all specific safety tests on the tools used in this activity, and be certified by the instructor to use the tools and equipment.

Activity Procedure
1. Cut pieces as per plans.
2. Nail 1 1/2″ edge to 4″ × 5 1/4″.
3. Sand joint flush with surface.

4. Clean all dust.

5. Apply cement to 1 1/2" edge and the 1 3/4" × 6" plastic laminate and let dry until you can touch it and it does not feel wet or stick to your skin. This is usually 10–15 minutes, depending on temperature, humidity, and brand of cement.

6. Carefully apply the 1 1/2" × 6" piece to the edge. Once the glued surfaces touch, it is stuck and cannot be moved.

7. Use a router with a flush trim bit to trim the laminate flush with all edges.

8. Use a sander to flush the laminate and the edge flush with the surface of the 4" × 6" piece. Anything out of square here will show through the surface laminate when applied.

9. Clean all dust from the surface.

10. Apply cement to both the surface 4" × 6" particleboard and the contact surface of the 4 1/2" × 6 1/2" laminate.

11. Let dry as before.

12. Carefully apply as before and press down with a J-roller.

13. Flush trim all edges with a flush trim bit.

14. Smooth the corner formed at the joint of the two laminates with 600 grit wet or dry silicon carbide sandpaper. This makes for a more comfortable finished corner.

Final Project Requirements

- Laminated project
- Follow-up questions

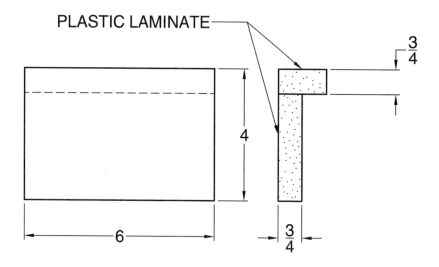

Name _____

Follow-Up Questions

1. What were your main challenges in completing this project?

   _____

   _____

   _____

2. Explain how plastic laminate is made.

   _____

   _____

   _____

3. Explain how contact cement works.

   _____

   _____

   _____

4. What kind of contact cement did you use?

   _____

Rubric for Applying Plastic Laminate
Student Guidelines

Construction   Possible points _____   Score _____

Did you ensure neatness, accuracy, dimensions, and correctness of views?

_____

Lamination   Possible points _____   Score _____

Did you complete a bill of materials, complete a list of resources, and figure total cost of materials?

_____

Laminate finishing   Possible points _____   Score _____

Did you keep a detailed list of steps to build the project? Did you keep accurate time records of estimated time and actual time for each step?

_____

Follow-up questions   Possible points _____   Score _____

Were all questions answered in a thoughtful manner?

_____

                                          Total possible points _____   Score _____

Name _____   Date _____   Course _____

SECTION PROJECT **5-1**

Surfacing Stock for Cabinet Face Frames

The two uses of the planer are preparing rails and stiles for the face frame of cabinets and preparing solid stock for raised panel door construction. In this activity, you will plane face frame material for a project to a precise thickness and width and evaluate the performance of the planer.

Resource Chapters

Chapter 25—Surfacing with Stationary Machines

Chapter 41—Frame and Panel Components

Objectives

After completing this project, you should be able to:

- Use the planer to prepare stock for use in face frame construction.
- Diagnose poor performance of a planer or jointer.
- Explain why it is most efficient to prepare rail and stile stock for several cabinets at one time.

Materials and Resources

- Solid stock, enough for face frame material for several cabinets
- Planer
- Table saw
- Calipers (preferably digital)

SAFETY NOTE
Before proceeding with this activity, you should have passed a general shop safety rule test, all specific safety tests on the tools used in this activity, and be certified by the instructor to use the tools and equipment.

Activity Procedure

1. Plane stock to a precise thickness needed for your rails and stiles, usually 3/4".

2. If one edge of the stock is not straight line ripped, it must be made straight on a jointer or use some other safe technique.

3. Rip the stock to 1/8" over the standard width that is used in your shop for rails and stiles.

4. Set the planer to cut the width of the material to within 1/16″ of the final width.
5. Turn the stock on edge and hold about 4–6 pieces together and run through the planer.
6. Repeat this process until all of your material is surfaced on one edge. This process removes saw marks.
7. Inspect the material to make sure the planer is adjusted properly. If not, make the proper adjustments as the text and planer owner's manual instruct.
8. Turn all the material over and repeat this process on the other edge, bringing the material to the exact width measurement.

Final Project Requirements

- Style and rail material finished to thickness and width ready for use in face frame cabinet construction
- Follow-up questions

Follow-Up Questions

1. What were your main challenges in completing this activity?

   _____

   _____

   _____

2. Why is the material for stiles and rails prepared in bulk?

   _____

   _____

   _____

3. How many total working hours did it take to prepare the stock from start to finish?

   Number of workers _____ × _____ hours = _____ working hours.

4. Did the planer perform perfectly? Explain your answer.

   _____

   _____

   _____

Name _____

Rubric for Surfacing Stock for Cabinet Face Frames
Student Guidelines

Jointer work Possible points_____ Score_____

Are the edges straight and square to face?

_____

Efficiency Possible points_____ Score_____

What were the total working hours?

_____

Planer work Possible points_____ Score_____

Did you properly evaluate planer performance and follow procedures?

_____

Follow-up questions Possible points_____ Score_____

Were all questions answered in a thoughtful manner?

_____

Total possible points _____ Score_____

Name _____ Date _____ Course _____

SECTION PROJECT **5-2**
Trinket Box

This is a beginner project in cabinetmaking. This project is a good practice for the beginning cabinet-maker because there are similarities to building a drawer.

Resource Chapter
Chapter 40—Case Construction

Objectives
After completing this project, you should be able to:
- Build a cabinetmaking project.
- Organize the work of cabinetmaking.
- Use a plan of procedure to keep the work on track.
- Use the resources in a typical cabinet shop to complete a cabinetmaking project.

Materials and Resources
- 2 1/2 board feet of solid wood
- 1/8″ 7 × 12 tempered hard board
- 3/4 × 3/4 brass butt hinge
- Box lid latch
- Sandpaper
- Stain
- Clear finish

SAFETY NOTE
Before proceeding with this activity, you should have passed a general shop safety rule test, all specific safety tests on the tools used in this activity, and be certified by the instructor to use the tools and equipment.

Activity Procedure
1. Study the prints and plan of procedure.
2. A study of Chapter 40 will give you some ideas of how to proceed with the construction of this project, especially Sections 40.3.6 to 40.4.1 and **Figure 40-9** and **Figure 40-10**.
3. Use the bill of materials to gather your materials.
4. Follow the procedure on the plan of procedure sheet provided.

Final Project Requirements

- Completed trinket box
- Follow-up questions
- Plan of procedure with updated time blocked in pencil
- Material cut sheet with checks as process was followed

Bill of Materials

Project Name: Trinket Box

Quantity	Units	Material Description	Price/Unit	Extended Price
2 1/2	BF	Solid wood, builder's choice		
1	each	1/8 inch thick, 7 x 12 tempered hard board		
1	pair	3/4 x 3/4 brass butt hinge		
1	each	Box lid latch		
1	sheet	Sandpaper		
		Stain		
		Clear finish		
			Subtotal	
			Tax	
			Total	

Name _____

Trinket Box Plan of Procedure

Steps		Hours 1-25
	Steps	
1	Study plans	
2	Fill out cut list	
3	Rough cut all materials	
	Sides	
4	Square a face and plane to thickness	
5	Square an edge and rip to width	
6	Cut drawer groove	
7	Cut miters on ends of side pieces	
8	Measure inside length of the groove on long and short sides	
9	For bottom, cut to above measurements minus 1/16	
10	Cut the feet shape on the bottom	
11	Apply glue & assemble sides together with bottom	
	Box Assembly	
12	Let dry overnight	
13	Square the top edge	
	Top	
14	Cut the top to 1/4" of size of the box	
15	Glue and clamp to the box assembly	
	Box Assembly	
16	Let dry overnight	
17	Flush trim or hand plane top flush with sides	
18	Sand with 120 grit garnet paper	
19	Cut top off with table saw	
	Box and Lid	
20	Mark lid and box so they stay oriented	
21	Square matting edges removing saw marks	
22	Lay out and mortise the hinge gain for butt hinges	
23	Install hinges and door latch	
24	Adjust until lid mates well with box	
25	Take hinges off for finishing process	
26	Remove any dents	
27	Putty cracks and holes	
28	Sand with 180 grit garnet paper	
29	Apply finish	
30	Reinstall hinges and box latch	
31	Gray is the suggested scheduling. Keep a record of the actual time scheduling by marking with pencil.	

Material Cut Sheet

Part Name	Material	Number Pieces	Rough Measurements			Cut Process Completed	Finish Measurements			Cut Process Completed
			Thickness	Width	Length		Thickness	Width	Length	

Name _____

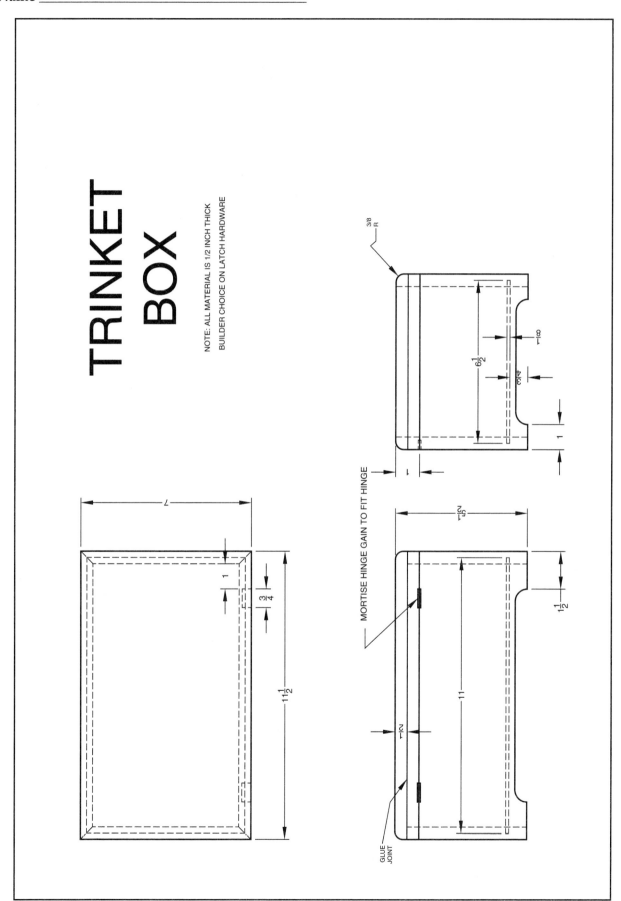

Follow-Up Questions

1. What were your main challenges in completing this project?
   _____
   _____

2. Explain how this project will be used.
   _____
   _____

3. Did you finish your project on time? What processes went faster than you thought they would, and what processes took longer?
   _____
   _____

4. Did you stay within your budget (this is what you figured on the bill of materials form) on the materials? What was the total cost of the materials for this project?
   _____
   _____

5. If you made $10.00 per hour, what was the cost of the labor on this project?
   _____

6. Explain the process that was used to mortise the hinge gains.
   _____
   _____

Name _____

Rubric for Trinket Box
Student Guidelines

Efficiency Possible points_____ Score_____

Did you keep accurate records on the plan of procedure for actual time?

_____

Craftsmanship Possible points_____ Score_____

Did you follow plans, sand out mill marks, and ensure corners are broken properly? Are edges flush at joints?

_____

Follow-up questions Possible points_____ Score_____

Were all questions answered in a thoughtful manner?

_____

Total possible points _____ Score_____

Name _____ Date _____ Course _____

SECTION PROJECT **5-3**
Bedside Face Frame Cabinets

This project is designed to give you practice in case construction with a face frame method of constructing cabinets. It features basic carcase construction basic frame techniques, recessed back, and simple panel door and drawer construction. The rail and stile joinery is not detailed in the plans because of the many different techniques. This is the cabinetmaker's choice. These techniques should be discussed and the best one for your shop chosen. There are also several options for the joinery in the drawer construction.

Resource Chapters

Chapter 9—Production Decisions

Chapter 22—Sawing with Hand and Portable Power Tools

Chapter 23—Sawing with Stationary Power Machines

Chapter 37—Joinery

Chapter 40—Case Construction

Chapter 41—Frame and Panel Components

Objectives

After completing this project, you should be able to:
- Master basic case construction skills.
- Make a panel cabinet door.
- Use European concealed hinges.
- Plan and execute the construction of a basic cabinet.
- Demonstrate skill in drawer making.
- Apply all the equipment and tools of the shop to the task of building cabinets.

Materials and Resources

- Solid 3/4″ stock for face frame
- Cabinet grade 3/4″ 4′ × 8′ MDF or plywood
- Drawer stock
- Drawer guides

- European concealed hinges
- 5 mm shelf pins
- Finish and sandpaper

SAFETY NOTE
Before proceeding with this activity, you should have passed a general shop safety rule test, all specific safety tests on the tools used in this activity, and be certified by the instructor to use the tools and equipment.

Activity Procedure

1. Study plans. Your instructor may want to make the stiles and rail material and cut the major parts of the cabinet as a class.
2. Make bill of materials and figure the estimated cost.
3. Make cut list.
4. The plans show the edges of the trim with a radius. Another option would be to add a shaped edge other than a radius. Discuss this with your instructor to find out what capabilities your shop has for shaping edges. Specify this in your plan of procedure.
5. Fill out the plan of procedure. Refer to Chapter 9, Sections 9.3 and 9.4.
6. Follow plan of procedure and keep time records on the plan of procedure form.
7. When making the drawer, make sure the drawer is exactly 1″ less than the distance between the stiles at the drawer opening. Most drawer glides require this; check manufacturers' installation instructions.

Final Project Requirements

- List of resources needed, tools needed, and bill of materials
- Plan of procedure
- Completed cabinet
- Follow-up questions

Name _____

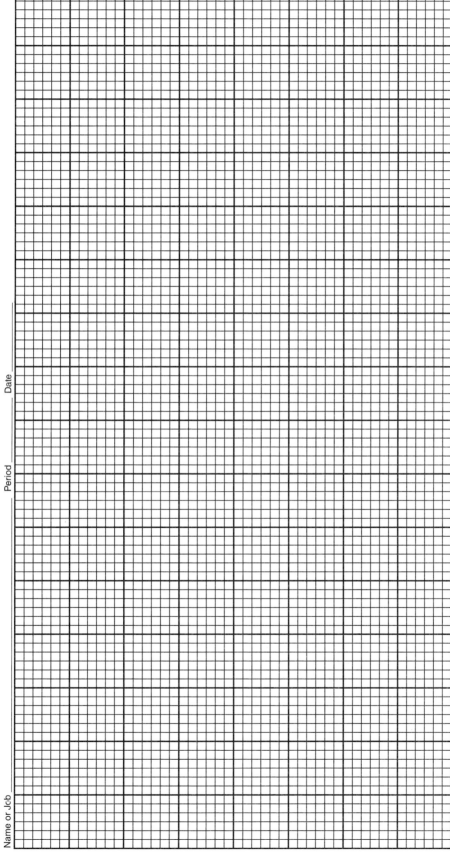
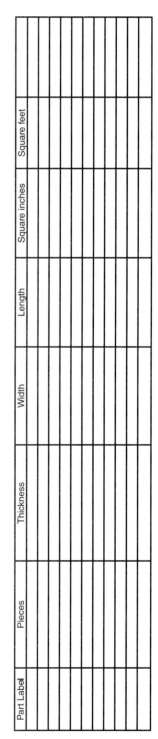

Bill of Materials

Project Name: _____

Quantity	Units	Material Description	Price/Unit	Extended Price

Subtotal	
Tax	
Total	

Name _____

Plan of Procedure

Chart with vertical axis "Steps" (1–32) and horizontal axis "Minutes, Hours, Days, Weeks, or Months" (1–25).

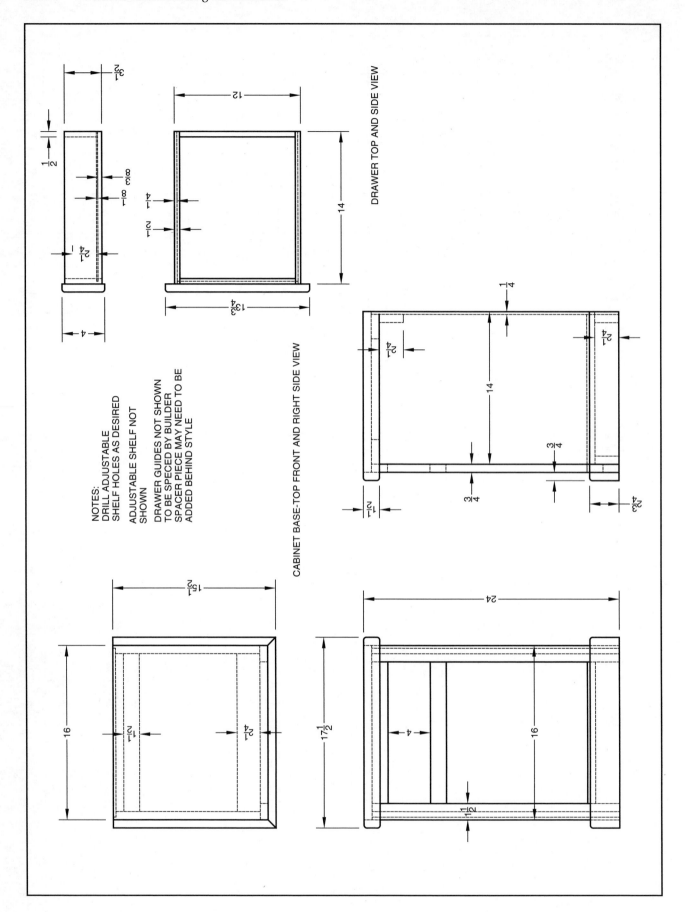

Name _____

Follow-Up Questions

1. What were your main challenges in completing this project?

   _____
   _____
   _____

2. Explain how this project will be used.

   _____
   _____
   _____

3. Did you finish your project on time? What processes went faster than you thought they would, and what processes took longer?

   _____
   _____
   _____

4. Did you stay within your budget (this is what you figured on the bill of materials form) on the materials? What was the total cost of the materials for this project?

   _____
   _____

5. How many hours did it take you to complete this project?

   _____
   _____

6. If you made $10.00 per hour, what was the cost of the labor on this project?

   _____

7. What technique did you use to complete your face frame? Explain why.

   _____
   _____

Copyright Goodheart-Willcox Co., Inc.
May not be reproduced or posted to a publicly accessible website.

Rubric for Bedside Face Frame Cabinets
Student Guidelines

Materials and resources Possible points_____ Score_____
Did you complete a bill of materials, fill out the cut list correctly, and figure total cost of materials?

_____

Process plan Possible points_____ Score_____
Did you keep a detailed list of steps to build the project? Did you keep accurate time records of estimated time and actual time for each step?

_____

Craftsmanship Possible points_____ Score_____
Did you follow plans, sand out mill marks, and ensure corners are broken properly? Are edges flush at joints? Are measurements accurate, and the cabinet square?

_____

Follow-up questions Possible points_____ Score_____
Were all questions answered in a thoughtful manner?

_____

Total possible points _____ Score_____

Name _____ Date _____ Course _____

SECTION PROJECT **5-4**

Wall Cabinets— Face Frame and Carcase Construction

This project is designed to give you practice in the basic face frame method of constructing cabinets. It features basic frame techniques, recessed back, and simple carcase construction. The rail and stile joinery is not detailed in the plans because of the many different techniques. A door is recommended but is not shown. The door may only cover the larger upper opening in the cabinet and leave the lower portion open. This is optional. There are so many ways to make a door, it is recommended that you study Chapter 43 and draw up your own design.

Resource Chapters

Chapter 9—Production Decisions

Chapter 22—Sawing with Hand and Portable Power Tools

Chapter 23—Sawing with Stationary Power Machines

Chapter 37—Joinery

Chapter 40—Case Construction

Chapter 41—Frame and Panel Components

Chapter 43—Doors

Objectives

After completing this project, you should be able to:

- Master basic carcase construction skills.
- Make cabinet doors.
- Use European concealed hinges.
- Plan and execute the construction of a basic cabinet.
- Use all the equipment and tools of the shop to the task of building cabinets.

Materials and Resources

- Cabinet grade 3/4" 4 × 8 MDF or plywood
- Solid wood stock for face frame
- 1/4" back material
- European concealed hinges and plates

- Finish supplies and sandpaper
- Shelf pins

SAFETY NOTE
Before proceeding with this activity, you should have passed a general shop safety rule test, all specific safety tests on the tools used in this activity, and be certified by the instructor to use the tools and equipment.

Activity Procedure

1. Study plans.
   A. Your instructor may want to make all the stiles and rail material and cut the major parts of the cabinet as a class.
   B. You will need to design your door before starting your plan of procedure.
2. Make bill of materials and figure the estimated cost.
3. Make cut list.
4. Before the plan of procedure can be completed, you and your instructor need to discuss what joints can be constructed in your shop. You will then need to choose the joints for the stile and rails and the cabinet sides.
5. Fill out plan of procedure. Refer to Chapter 9, Sections 9.3 and 9.4.
6. You may want to discuss adding shaped edges. Discuss with your instructor the capabilities of your shop to do this. Specify this in your plan of procedure.
7. Follow plan of procedure and keep time records.

Final Project Requirements

- Bill of materials
- Cut list form
- Plan of procedure
- Shop drawings of door design
- Completed wall cabinet
- Follow-up questions

Name _____

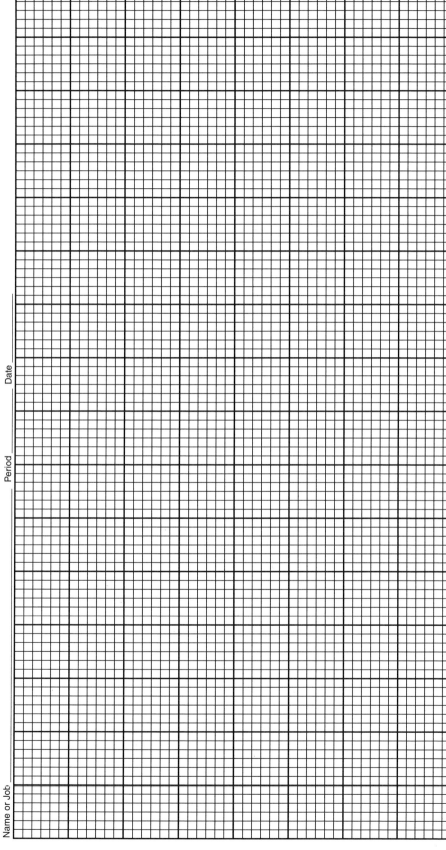

Bill of Materials

Project Name: _____

Quantity	Units	Material Description	Price/Unit	Extended Price
			Subtotal	
			Tax	
			Total	

Name _____

Material Cut Sheet

Part Name	Material	Number Pieces	Rough Measurements			Cut Process Completed	Finish Measurements			Cut Process Completed
			Thickness	Width	Length		Thickness	Width	Length	

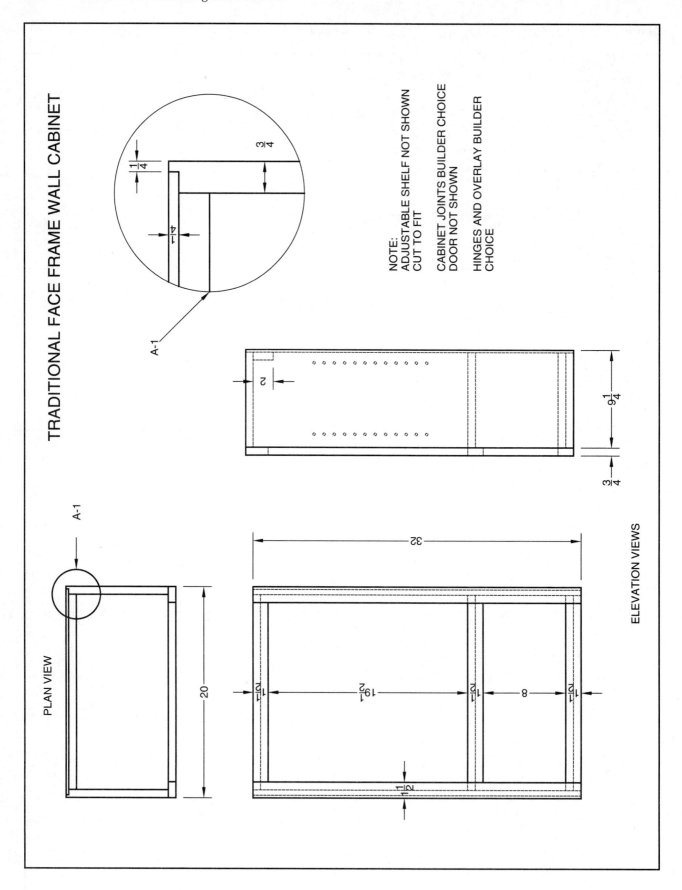

Name _____

Follow-Up Questions

1. What were your main challenges in completing this project?

   _____

   _____

2. Explain what a face frame is.

   _____

   _____

3. Did you finish your project on time? What processes went faster than you thought they would, and what processes took longer?

   _____

   _____

   _____

4. Did you stay within your budget (this is what you figured on the bill of materials form) on the materials? What was the total cost of the materials for this project?

   _____

   _____

5. If you made $10.00 per hour, what was the cost of the labor on this project?

   _____

6. What style door did you choose for your cabinet?

   _____

Rubric for Wall Cabinets—Face Frame and Carcase Construction
Student Guidelines

Materials and resources Possible points_____ Score_____

Did you complete a bill of materials, complete a list of resources, and figure total cost of materials?

_____

Process plan Possible points_____ Score_____

Did you keep a detailed list of steps to build the project? Did you keep accurate time records of estimated time and actual time for each step?

_____

Craftsmanship Possible points_____ Score_____

Did you follow plans, sand out mill marks, and ensure corners are broken properly? Are edges flush at joints? Are measurements accurate, and the cabinet square?

_____

Follow-up questions Possible points_____ Score_____

Were all questions answered in a thoughtful manner?

_____

Total possible points _____ Score_____

Name _____ Date _____ Course _____

SECTION PROJECT **5-5**

Frameless Bedside Stand Cabinets

This project is designed to give you practice in the basic frameless method of constructing cabinets. It will feature a solid full overlay mount door and recessed back. There are options for door design and drawer construction. Deviation from the plans should be backed up by your shop drawings. Another optional challenge is to modify the cabinet for the 32mm System.

Resource Chapters

Chapter 9—Production Decisions

Chapter 22—Sawing with Hand and Portable Power Tools

Chapter 23—Sawing with Stationary Power Machines

Chapter 34—Overlaying and Inlaying Veneer

Chapter 37—Joinery

Chapter 40—Case Construction

Chapter 41—Frame and Panel Components

Chapter 43—Doors

Objectives

After completing this project, you should be able to:

- Demonstrate skill in carcase construction.
- Master edgebanding techniques.
- Install European concealed hinges.
- Cut cabinet parts from 4′ × 8′ sheets.
- Construct drawers.
- Use equipment and tools of the shop to the task of building cabinets.

Materials and Resources

- Cabinet grade 3/4″ 4′ × 8′ MDF or plywood
- Edgeband material
- Drawer stock

- Drawer guides
- European concealed hinges
- Finish supplies and sandpaper
- Shelf pins

SAFETY NOTE
Before proceeding with this activity, you should have passed a general shop safety rule test, all specific safety tests on the tools used in this activity, and be certified by the instructor to use the tools and equipment.

Activity Procedure

1. Study the plans.
2. Make a cut list.
3. Make a bill of materials.
4. Make a list of resources needed other than materials, such as tools and equipment.
5. Before making the plan of procedure, a decision must be made regarding the type of joints, edgeband and drawer glides to be used. Your shop capabilities and instructor's preference will influence these decisions.
6. Make a plan of procedure. Refer to Chapter 9, Sections 9.3 and 9.4.
7. Execute your plan of procedure.

Final Project Requirements

- Bill of materials
- Cut list form
- Plan of procedure
- Shop sketches of door design
- Finished frameless bedside stand
- Follow-up questions

Name _____

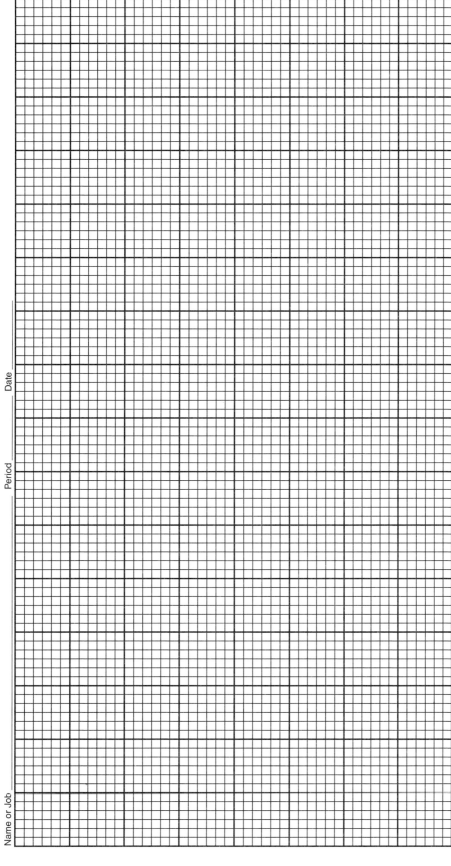
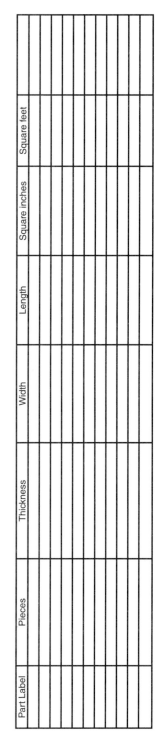

Bill of Materials

Project Name: _____

Quantity	Units	Material Description	Price/Unit	Extended Price
			Subtotal	
			Tax	
			Total	

Name _____

Material Cut Sheet

Part Name	Material	Number Pieces	Rough Measurements			Cut Process Completed	Finish Measurements			Cut Process Completed
			Thickness	Width	Length		Thickness	Width	Length	

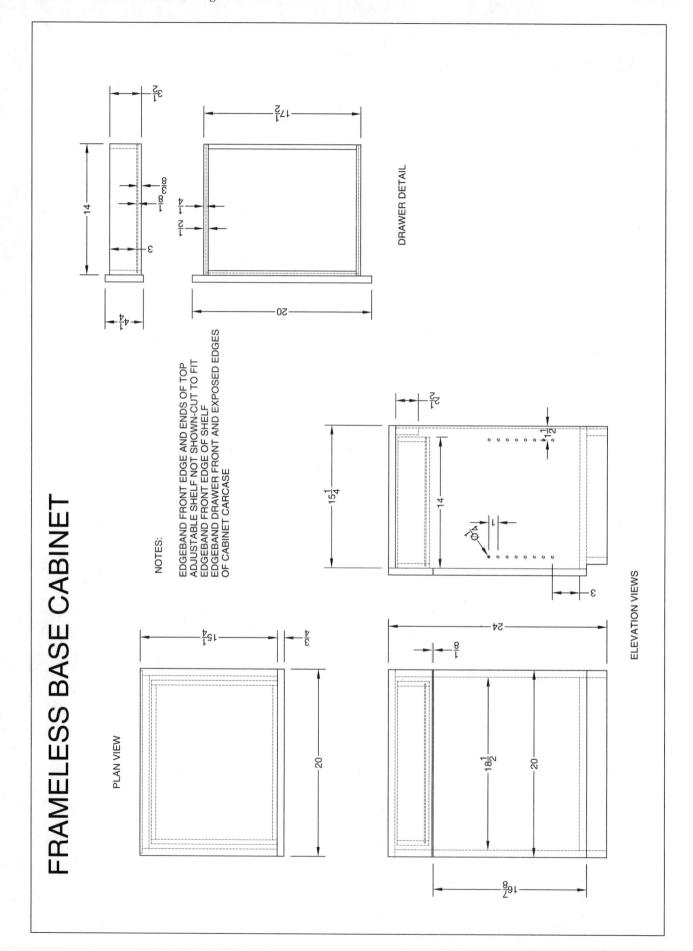

Name _____

Follow-Up Questions

1. What were your main challenges in completing this project?

   _____

   _____

2. Explain how this project will be used.

   _____

   _____

3. Did you finish your project on time? What processes went faster than you thought they would, and what processes took longer?

   _____

   _____

   _____

4. Did you stay within your budget (this is what you figured on the bill of materials form) on the materials? What was the total cost of the materials for this project?

   _____

   _____

5. If you made $10.00 per hour, what was the cost of the labor on this project?

   _____

6. Did you deviate from the plans? Explain your answer.

   _____

   _____

Rubric for Frameless Bedside Stand Cabinets
Student Guidelines

Process plan   Possible points_____   Score_____

Did you ensure neatness, accuracy, dimensions, and correctness of views?

_____

Materials and resources   Possible points_____   Score_____

Did you complete a bill of materials, complete a list of resources, and figure total cost of materials?

_____

Craftsmanship   Possible points_____   Score_____

Did you follow plans, sand out mill marks, and ensure corners are broken properly? Are edges flush at joints? Are measurements accurate, and the cabinet square?

_____

Follow-up questions   Possible points_____   Score_____

Were all questions answered in a thoughtful manner?

_____

                    Total possible points _____   Score_____

Name _____ Date _____ Course _____

SECTION PROJECT 5-6
Frameless Wall Cabinets

This project is designed to give you practice in the basic frameless method of constructing cabinets. It will feature a solid flush mount door, recessed back. Butt joints can be biscuit joined, doweled, or be finish nailed.

Resource Chapters

Chapter 9—Production Decisions

Chapter 22—Sawing with Hand and Portable Power Tools

Chapter 23—Sawing with Stationary Power Machines

Chapter 34—Overlaying and Inlaying Veneer

Chapter 37—Joinery

Chapter 40—Case Construction

Chapter 41—Frame and Panel Components

Chapter 43—Doors

Objectives

After completing this project, you should be able to:

- Explain the procedure to construct a simple frameless European-style cabinet.
- Compare the work and expertise needed to build a frameless cabinet to building a face frame cabinet.

Materials and Resources

- Cabinet grade 3/4" and 1/4" panel materials
- Suitable edgebanding
- European concealed hinges with full overlay plates
- Finish supplies and sandpaper
- 5 mm shelf pins

SAFETY NOTE
Before proceeding with this activity, you should have passed a general shop safety rule test, all specific safety tests on the tools used in this activity, and be certified by the instructor to use the tools and equipment.

Activity Procedure

1. Study the plans.
2. Make a cut list.
3. Make a bill of materials.
4. Make a list of resources needed other than materials, such as tools and equipment.
5. Before making the plan of procedure, a decision must be made regarding the type of joints and edgeband to use. Shop capabilities and your instructor's preference will influence these decisions.
6. This project could be used to practice applying plastic laminate or veneer. Discuss these possibilities with your instructor before filling out the plan of procedure.
7. Make a plan of procedure. Refer to Chapter 9, Sections 9.3 and 9.4.
8. Execute your plan of procedure.

Final Project Requirements

- List of resources needed, tools needed, and bill of materials
- Plan of procedure
- Finished frameless wall cabinet
- Follow-up questions

Name _____

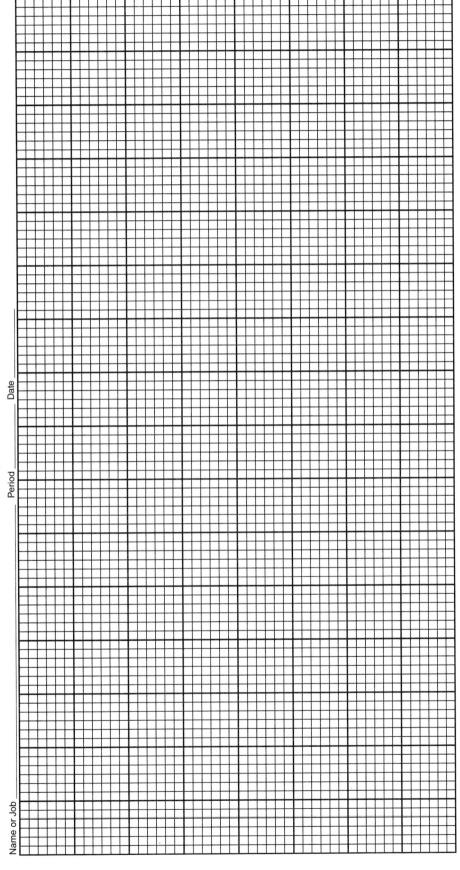

Bill of Materials

Project Name: _____

Quantity	Units	Material Description	Price/Unit	Extended Price

Subtotal	
Tax	
Total	

Name _____

Material Cut Sheet

Part Name	Material	Number Pieces	Rough Measurements			Cut Process Completed	Finish Measurements			Cut Process Completed
			Thickness	Width	Length		Thickness	Width	Length	

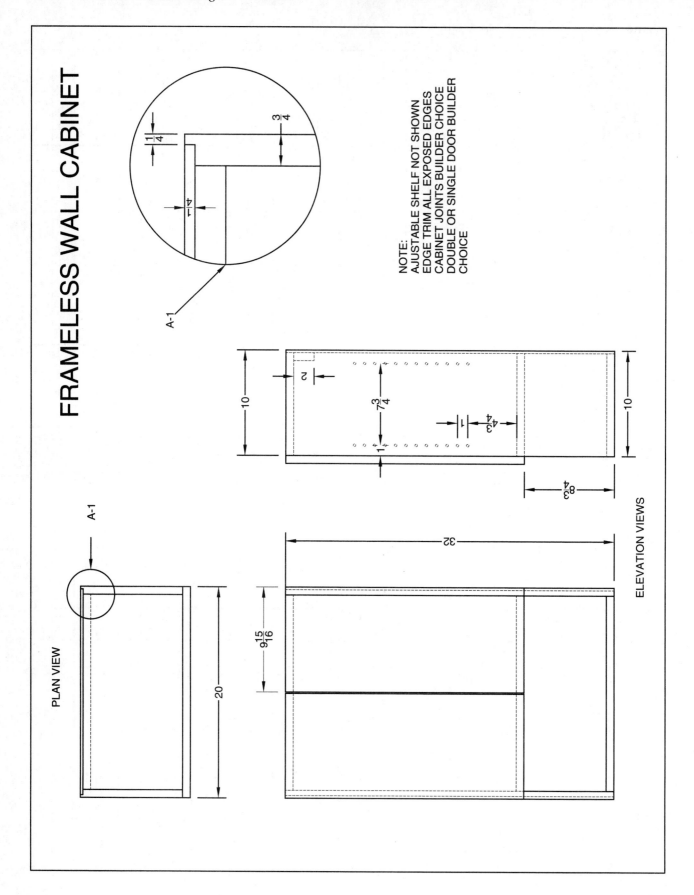

Name _____

Follow-Up Questions

1. What were your main challenges in completing this project?

   _____

   _____

   _____

2. Explain how this project will be used.

   _____

   _____

   _____

3. Did you finish your project on time? What processes went faster than you thought they would, and what processes took longer?

   _____

   _____

   _____

4. If you made $10.00 per hour, what was the cost of this project?

   _____

5. Did you stay within your budget (this is what you figured on the bill of materials form) on the materials? What was the total cost of the materials for this project?

   _____

   _____

Rubric for Frameless Wall Cabinets
Student Guidelines

Materials and resources   Possible points_____ Score_____

Did you complete a bill of materials, complete a list of resources, and figure total cost of materials?

_____

Process plan   Possible points_____ Score_____

Did you keep a detailed list of steps to build the project? Did you keep accurate time records of estimated time and actual time for each step?

_____

Craftsmanship   Possible points_____ Score_____

Did you follow plans, sand out mill marks, and ensure corners are broken properly? Are edges flush at joints? Are measurements accurate, and the cabinet square?

_____

Follow-up questions   Possible points_____ Score_____

Were all questions answered in a thoughtful manner?

_____

                                           Total possible points _____ Score_____

Name _____ Date _____ Course _____

SECTION PROJECT 6-1
Fixing a Dent

While cabinetmakers strive not to make unnecessary scratches and dents in their projects, sometimes mistakes happen. Therefore, the cabinetmaker should be skilled at fixing these mistakes.

Resource Chapter
Chapter 50—Preparing Surfaces for Finish

Objective
After completing this project, you should be able to:
- Use a household iron to remove a dent.

Materials and Resources
- Household iron
- 4″ × 4″ wood scraps
- 120 grit sandpaper

SAFETY NOTE
Before proceeding with this activity, you should have passed a general shop safety rule test, all specific safety tests on the tools used in this activity, and be certified by the instructor to use the tools and equipment.

Activity Procedure
1. Find a small piece of soft maple, pine, or poplar and cut to 4″ × 4″.
2. Use a bell-faced hammer to dent the wood.
3. Place a 16d nail on the wood and hit the nail to put a small dent in the wood.
4. In both cases, try to not cut the fibers, but only create a dent.
5. Soak the dented areas with warm water.
6. Being careful not to touch the hot surface or lay it next to flammable materials, place the iron on the dent. The water should steam the dent out.
7. Repeat the process if your first attempt does not remove the dent.
8. Lightly sand the piece to determine if the dent completely disappears.

Final Project Requirements

- Wood scrap with dent removed

Follow-Up Questions

1. What did you learn from this activity?

   _____
   _____
   _____
   _____

2. Did you burn the wood with the iron? If so, how do you think you can avoid this from happening the next time?

   _____
   _____
   _____
   _____

3. At what point in the cabinetmaking process do you think dent removal should take place?

   _____
   _____
   _____
   _____

Name _____

Rubric for Fixing a Dent
Student Guidelines

Dent removal Possible points _____ Score _____

Did you successfully remove the dent?

_____

Follow-up questions Possible points _____ Score _____

Were all questions answered in a thoughtful manner?

_____

Total possible points _____ Score _____

Name _____ Date _____ Course _____

SECTION PROJECT 6-2
Finish Sample Set

The main function of finish is to protect and beautify a wood or metal product. The purpose of this activity is to become familiar with the many finishes used in your shop. In this activity, you will make a sample set of those finishes.

Resource Chapters

Chapter 49—Finishing Decisions

Chapter 50—Preparing Surfaces for Finish

Chapter 51—Finishing Tools and Equipment

Chapter 52—Stains, Fillers, Sealers, and Decorative Finishes

Chapter 53—Topcoatings

Objective

After completing this project, you should be able to:

- Demonstrate an understanding of the finishing methods used in your shop.

Materials and Resources

- 1/4″ solid wood or plywood
- Finishing supplies used to finish cabinets in your shop
- Plan of procedure sheet

SAFETY NOTE
- Make sure when applying solvent-based finishes to work in a well-ventilated area and to use proper respirators.
- Before proceeding with this activity, you should have passed a general shop safety rule test, all specific safety tests on the tools used in this activity, and be certified by the instructor to use the tools and equipment.

Activity Procedure

1. Cut three to six 1/4″ × 3″ × 6″ wood samples (one sample of each wood for each finishing recipe).

2. Cut a 1/16″ deep saw kerf across the grain of the pieces, 2″ from the end.

Copyright Goodheart-Willcox Co., Inc.
May not be reproduced or posted to a publicly accessible website.

3. Finish sand each piece.
4. Write your finishing procedures in the plan of procedure sheet.
5. Place a piece of masking tape on the edge of the saw kerf to mask off the 2″ side.
6. Perform your finishing procedures on the 4″ portion of the piece, leaving the 2″ portion unfinished.
7. Repeat for each sample.

Final Project Requirements

- Finished samples
- Plan of procedure sheet with finishing procedure (time section can be left blank)

Name _____

Plan of Procedure

Minutes, Hours, Days, Weeks, or Months: 1, 2, 3, 4, 5, 6, 7, 8, 9, 10, 11, 12, 13, 14, 15, 16, 17, 18, 19, 20, 21, 22, 23, 24, 25

Steps: 1, 2, 3, 4, 5, 6, 7, 8, 9, 10, 11, 12, 13, 14, 15, 16, 17, 18, 19, 20, 21, 22, 23, 24, 25, 26, 27, 28, 29, 30, 31, 32

Follow-Up Questions

1. What were your main challenges in completing this project?

   _____
   _____
   _____
   _____

2. Explain how this project will be used.

   _____
   _____
   _____
   _____

3. What thinners were used in your finishing procedures?

   _____
   _____
   _____
   _____

4. What methods did you use to apply your finishes? Explain.

   _____
   _____
   _____
   _____

Name _____

Rubric for Finish Sample Set
Student Guidelines

Required number of samples Possible points_____ Score_____

Did you complete the number of samples set by the instructor?

_____

Followed directions Possible points_____ Score_____

Did you follow correct procedures for removing and storing supplies and for cleanup? Were rags disposed of properly and brushes and equipment cleaned?

_____

Plan of procedure Possible points_____ Score_____

Did you fill out the sheet in detail?

_____

Craftsmanship and neatness Possible points_____ Score_____

Were the finishes executed correctly?

_____

Follow-up questions Possible points_____ Score_____

Were all questions answered in a thoughtful manner?

_____

Total possible points _____ Score_____

Copyright Goodheart-Willcox Co., Inc.
May not be reproduced or posted to a publicly accessible website.